GEOGRAPHY'S QUANTITATIVE REVOLUTIONS

GEOGRAPHY'S

EDWARD A. ACKERMAN AND THE

QUANTITATIVE

COLD WAR ORIGINS OF BIG DATA

REVOLUTIONS

ELVIN WYLY

WEST VIRGINIA UNIVERSITY PRESS
MORGANTOWN 2019

ISBN
Cloth 978-1-949199-08-6
Paper 978-1-949199-09-3
Ebook 978-1-949199-10-9

Library of Congress Cataloging-in-Publication Data
Names: Wyly, Elvin K., author.
Title: Geography's quantitative revolutions : Edward A. Ackerman and the
    Cold War origins of big data / Elvin Wyly.
Description: First edition. | Morgantown : West Virginia University Press,
    2019. | Includes bibliographical references and index.
Identifiers: LCCN 2019010618 | ISBN 9781949199086 (cloth) | ISBN
    9781949199093 (paperback) | ISBN 9781949199109 (ebook)
Subjects: LCSH: Ackerman, Edward A. (Edward Augustus), 1911–1973. |
    Geographers–United States–Biography. | Geography–History–20th
    century. | Geography–Philosophy. | Geography–Methodology. |
    Geographic information systems–History. | Neo-Kantianism–History–
    20th century.
Classification: LCC G69.A35 W95 2019 | DDC 910.72/02 [B]–dc23
LC record available at https://lccn.loc.gov/2019010618

Book and cover design by Than Saffel / WVU Press

In loving memory of Robert S. Wyly,
1920–2018

Contents

Preface

In a period of vibrant innovation in the decades after the Second World War, a "quantitative revolution" swept through many of the social sciences, altering the logics, methods, and practices of knowledge production. This revolution was particularly transformative in the field of human geography, in part because of the discipline's aspirations to overcome its small size and weak position in the pecking order of elite universities—particularly in the United States. The revolution was also enmeshed with the development of science and technology practices that came to be known as the military-industrial complex. Yet geography's revolution was remarkably short-lived, quickly followed by a series of dramatic insurgencies questioning its underlying philosophy, methods, and politics. In the space of little more than a decade—from the early 1960s to the early 1970s—the field became a dynamic, dialectical site of struggle between positivist spatial science and an evolving plurality of alternatives and oppositional movements. While many of the earliest challenges came from comparatively conservative currents in regional and cultural geography, the field was soon remade by Marxist, environmentalist, antiracist, and feminist mobilizations.

A great deal has been written about geography's quantitative revolution. Much of that literature has emphasized the dramatic turning points of the late 1960s and early 1970s, and for many observers today—conservatives and radicals alike—contemporary history begins with the intellectual and political evolution of David Harvey. His 1969 *Explanation in Geography*, written in his years at Bristol, became an instant classic, widely described as the Bible of a quantitative positivist geography that had by then become the mainstream. Even as he submitted galley proofs of *Explanation*, however, he was rethinking everything after arriving to take up a position at Johns Hopkins. Baltimore was in flames with antiracist and antiwar protests, and Harvey's radicalization eventually culminated in a sort of liberation theology for the discipline's scientific scriptures in the deeply influential 1973 Marxist manifesto *Social Justice and the City*. Harvey also engaged in high-profile battles with the field's most prominent defender of spatial science, Brian J. L. Berry.

Today, however, our understanding of geography's quantitative revolution is itself in the midst of a wide-ranging reconsideration. Three factors are most significant. First, we are seeing further into the prehistory of the field's revolutionary moments. Thanks to the genealogical and oral history work led over the past two decades by Trevor Barnes, we are gaining a much more situated, intimate biographical understanding of the establishment geography that was the target of revolutionary challenges in the 1960s and 1970s. We now know much more about the extent of geographers' roles in the intelligence apparatus constructed by the United States in the 1940s and in some of the Nazis' wartime operations. Thanks to Neil Smith's geopolitical and biographical history of Isaiah Bowman, we know more about the militarization of geographical knowledge in the First World War, the discipline's entrenched reliance on the eighteenth-century philosophy of Immanuel Kant, and the political struggles involved in the catastrophic decision to close Harvard's geography program in the 1940s. Second, it has become clear that we are now in the midst of a new and more powerful quantitative revolution, with a complex mixture of continuities and ruptures from the paradigmatic struggles of previous generations. As a formalized discipline, geography remains a complex, pluralist field shaped by enduring dichotomies between physical and human subjects, descriptive versus explanatory modes of thought, and mainstream/conservative opposed to insurgent and radical political projects. Yet beyond the formal confines of the academy, geography has suddenly achieved a nearly universal popularity—as the formalized infrastructures of geographical information systems (GIS) diffuse through governments and businesses and as military technologies like the Global Positioning System (GPS) fuse with the smartphone "mobile revolution" and big data in the Silicon Valley "disruptions" of everyday life. Everyone seems to love the geography that is shaping the spaces and places of their daily connected lives, even as many continue to see the formal field of geography as little more than the memorization of locational trivia. Do we need geographers? Aren't there plenty of apps for that?

Put simply, the quantitative revolution is both older and newer than we once thought. It is essential that we recover lost memories from the work of previous generations of human geographers *and* that we examine the explicit and hidden connections to the latest frontiers of the observation, measurement, and manipulation of information about spaces and places. Our sense of history and geography— always a domain of contestation in science and philosophy— is undergoing an episode of profound, accelerated, and nonlinear evolution. Geography, at least as understood as the engagement with information about the spaces and places of people and environments in a diverse world, is more popular than ever before. And yet one of the paradoxes of our age of algorithmic efficiency and artificial intelligence is that it is no longer clear how many humans we need who call themselves geographers.

In this short book, I wrestle with this paradox of geography's popularity in an age of automation and dehumanization. My approach is biographical: I examine the interplay of past and present theories and technologies through a close examination of the work of a neglected, often-overlooked figure in the discipline's evolution: Edward A. Ackerman (1911–1973). In September 1963, Ackerman, who had been invited to be honorary president of the Association of American Geographers, delivered a presidential address titled "Where Is a Research Frontier?" Ackerman's address was a panoramic survey of the advances achieved by the mathematical and physical sciences in the first half of the twentieth century and a bold exhortation for a weak, marginalized, obsolete geography to step up its game, to work harder to reach and advance those frontiers. Ackerman argued for a deeper engagement with the quantitative methods that were already becoming so important in the field. But he situated quantitative methods within a much more ambitious explanatory framework called General Systems Theory. He portrayed a "revolution of rationalism" in the economic structure of the United States. He described the role of science and technology in the "social problem of automation." He spoke of how "cybernation" was eliminating the need for individual human decision-making, changing the operations of the nation's defense program, and enabling breakthroughs in the understanding of "the process of human thought itself." These futuristic technological themes, of course, are all part of the Cold War infrastructure that John Bellamy Foster and Robert McChesney have called "surveillance capitalism,"[1] which has now developed into the transnational circuits of cloud computing and National Security Agency (NSA) "collect it all" monitoring doctrines.

My central argument is that we can gain a much better understanding of quantitative revolutions—today's and yesterday's—by undertaking a close study of the thought, life, and context of Ackerman. The *content* of his 1963 presidential address illuminates an optimistic, forward-looking faith in modernist, linear scientific progress and the inevitable triumph of an American Dream through the perilous times of wars hot and cold. The unstated *context* of his address reveals a more situated, happenstance, and bittersweet biographical trajectory that positioned him between academic geography and the military-industrial complex. Ackerman had been intimately involved in the painful disaster of the Harvard closure, and he played a central role in the applied geography practiced in the Office of Strategic Services (OSS), the forerunner of today's CIA. He was also the product of an older mode of geographical thought that looked to the past for its legitimation, back to Kant and to the nineteenth-century fusion of geology and evolutionary theory that created the foundations for American human geography. After the war and Harvard geography's demise, Ackerman accepted a series of positions—consultant to MacArthur's occupation forces in Japan, chief geographer for the President's Water Resources Policy Commission, assistant general manager of the Tennessee Valley Authority, and then finally executive officer at the Carnegie Institution of Washington—that put him at the center of the emerging think-tank infrastructure of American science and technology policy.

By the time he became honorary president for the Association of American Geographers, then, Ackerman had

developed a truly unique intellectual and personal, embodied perspective shaped by geography's historical reverence
toward a slow, evolutionary past and its sudden ambitions to
catch up to the futuristic advancing wave of mathematical,
positivist big science. Situated in the ferment of the postwar
fascination with quantitative behavioral science and cybernetics, Ackerman's 1963 address helps us to see important
continuities between today's worlds of big data and the eigh-
teenth- and nineteenth-century philosophies used to build
knowledge of a diverse world of different peoples and natural environments. His thought and work in the early phases
of the computer revolution in social research, moreover, help
us to see some of the early paradoxes of the extraordinary
transformation in the nature of scale, aggregation, and statistical inference in the practices of what Donna Haraway
has diagnosed as "technoscience."[2] Finally, a consideration
of Ackerman forces us to confront the poignant essence of
humanity in the practice of scholarly inquiry. Orphaned as
a young child, Ackerman landed a scholarship at Harvard
and soon became recognized as one of the most brilliant
geographers of his generation. He was kind, thoughtful, and
principled. And yet in his struggles to build his own career
and to nurture an emergent geographical science, Ackerman
was shaped by and contributed to a military-industrial complex premised upon hierarchical control and the management
of violence. A consideration of the situated, human contradictions of Ackerman is important as we are forced to confront
today's increasingly automated geography, where algorithmic
advances are rapidly transforming the meanings of individual human choices, constraints, and perceptions—and
the meanings of human responsibility, care, and empathy. Algorithmic aggregation through crowdsourcing, the

adaptive auto-recommend interfaces augmented by artificial intelligence, and the pattern-recognition data-mining techniques now widely used in both corporate and military surveillance and micro-targeting—all of these cybernetic advances represent the culmination of trends foreseen by Ackerman more than half a century ago. All of these advances are quickly dehumanizing geographical thought and practice. Algorithmic geographic thought is always changing, and indeed it has become very explicitly evolutionary in the world that the science historian George Dyson calls the "universe of self-replicating code."[3] But you and I—as human author and reader—cannot really talk, persuade, and fight with the algorithms of surveillance capitalism. Previous generations of authors and readers talked, taught, and learned in university seminar rooms and street protests—struggling over the meanings of science, justice, and "progress." While this still happens today, it is increasingly mediated by the evolutionary algorithmic adaptations of a cybernetic infrastructure that is redefining human agency and human responsibility.

This is a work of theory and synthesis, easily assembled from three types of sources. First, we can evaluate Ackerman's position within the history of formalized ideas in the discipline through the published literature of the evolving, contested canon of American geographical thought. Second, the Edward A. Ackerman Papers at the American Heritage Center at the University of Wyoming offer a treasure trove of archived documents and correspondence; these files offer evocative clues to the biography of Ackerman's experiences and ideas and his role in the early years of the field's quantitative revolution. Third, we can analyze the connections between Ackerman's thought and the military-industrial complex (yesterday and today) by using the simple "open source intelligence" methods he helped refine at the OSS in the 1940s—careful scrutiny of contemporary public-interest journalism and other readily available government and scholarly documents.

It has been more than half a century since the quantitative revolution began to transform geography, and human memories are fading fast. This era is now widely recalled as an epoch of austere mathematics and bold ambitions for geography to become a "true" science on the advancing frontiers of positivist observation of the external world. This is an important part of the story, but it is partial and incomplete. Ackerman's life and legacy remind us of a hidden history of positivist geographical thought—a blend of cybernetic engineering metaphors and a distorted form of idealist phenomenology that was hijacked by American military hegemony and the earliest epistemological strains of neoliberalism. The result, a hybrid that I call "militant neo-Kantianism," has corrupted the discipline and accelerated the algorithmic evolutionary dehumanization of human geography. We must understand this history—a history that is alternately forgotten, distorted, and suppressed—so that we can decide what kind of human geographers we wish to become, and what kinds of human and nonhuman worlds are possible and worth fighting to build.

Acknowledgments

Thank you for reading these words. Have you fallen in love, as have I, with the joy of reading acknowledgments even when you'll never have time to read the whole book? Even if the book's not worth reading, the acknowledgments always are.

So here's what you need to know. A few years ago, I was reading an astonishing, brilliant article by Trevor Barnes. I can't remember which one: Trevor thinks and writes so much faster than I can read. I noticed a name, vaguely familiar, in Trevor's narrative of some of the most important historical moments of geography's involvement in the U.S. military-industrial complex. Ackerman. Distant memories stirred of seeing that name in various footnotes from years ago, in stories of geography's quantitative revolution, and an ambitious but short-lived fad known as General Systems Theory. I grabbed an old copy of the *Annals of the Association of American Geographers* and found the 1963 article that had, for a few years, been so widely cited: "Where Is a Research Frontier?" It was from a talk delivered by Edward A. Ackerman, executive officer of the Carnegie Institution of Washington and honorary president of the AAG. In another article, Trevor mentioned that Ackerman's papers were archived at the American Heritage Center at the University of Wyoming.

I read about these connections in early 2016, as we were planning family and sabbatical research travel, in the months when U.S. politics was delivering ever more bizarre lessons in the automated epistemologies of Twitterbots, viral "dark posts" on Facebook, and the weaponized racist memes of a cybernetic "alt-right." As we were making travel arrangements, in the background I was running code to mine the YouTube API to watch the audience formation processes as the Republican National Convention was livecast from Cleveland. Angry voices blared from the television in the other room. Retired lieutenant general Michael Flynn, once head of the Defense Intelligence Agency and architect of the Joint Special Operations Command's "We Track 'Em, You Whack 'Em" integration of cell phone geolocation and drone-strike assassinations, led the crowd in chants against Hillary Clinton: "LOCK HER UP!" PayPal cofounder and Facebook first investor Peter Thiel praised Donald Trump as a "builder" who could "rebuild America" to the innovation glory of the 1960s—those were the research frontiers of Ackerman's 1963 vision—when "all of America was high tech," not just Silicon Valley.

Our plan was to drive to Winnipeg on the first leg of an extended itinerary.

Laramie, Wyoming, is on the way to Winnipeg.

I didn't get as much time in the Ackerman archives as I'd hoped. A transmission catastrophe killed our trusty 2001 Nissan Sentra on a long uphill climb in the Blue Mountains in eastern Oregon, leading to a few adventures that delayed our journey. By this point, however, I was seeing connections everywhere. Ackerman had described a process of "cybernation" in 1963 that is today manifest in the fully automated directory assistance chatbot that refused—no matter how many loud profanities I yelled into the cell phone as we stood on the side of the road—to connect to a human operator who could give us the number for a tow truck. By the time we finally got to a junkyard and repair shop, wandering through a gravel parking lot in search of the office front door, the proprietor smiled and chuckled. He thought we were part of the endless stream of people who had been showing up in recent weeks, searching for the augmented-reality Pokémon characters that had somehow been coded in his parking lot. Eventually we got to Wyoming, and a short but exhilarating immersion in the Ackerman archives made it clear that we're now in a strange parallax view of the history and present condition of human geography. We're living in a strangely mutated version of the technologically advanced world that Ackerman glimpsed more than half a century ago. Geography is more popular than ever before, as billions of people with powerful mobile computers in their pockets navigate a planet of networked informational spaces and places. And yet all the advances in algorithmic automation and artificial intelligence present uncomfortable questions: why do we need "human geographers," or a discipline called "geography," when everyone has a smartphone with advanced geospatial capacities and direct connections to globally networked human communication and intelligence?

In wrestling with this question, I am grateful to many people. Thanks to Trevor for the inspiration. Thanks to Peter Wissoker for generous and wise counsel that led me to Derek Krissoff at West Virginia University Press. Derek has been kind and patient in his support for this project; I am grateful for the time he allowed so that I could revise a first draft of the manuscript. And revision was indeed necessary in this short but complex project. This book is part biography of Edward Ackerman and part techno-cultural analysis of our present accelerating moment of robotic geographies. Audrey Kobayashi, Joel Wainwright, and John Pickles read through the entire manuscript and offered detailed comments, questions, and recommendations. If there's anything worth reading here, the credit goes to them for their thoughtful reviews and guidance, as well as other currents of inspiration from, inter alia, John's brilliant *Ground Truth: The Social Implications of Geographical Information Systems*, Audrey's powerful *Urban Geography* article "Neoclassical Urban Theory and the Study of Racism in Geography," and Joel's devastating *Geopiracy: Oaxaca, Militant Empiricism, and Geographical Thought*. Audrey, Joel, and John's careful readings and recommendations took me further out on the edges of multiple genealogical and theoretical frontiers, helping me rethink and rewrite key parts of the story. Brian Berry also offered several valuable recollections, clarifications, and reflections. Thanks also to an evolving constellation of students whose rigor, eloquence, and creativity inspire and energize: Joseph Daniels, Emily Rosenman, Sage Ponder, Luke Barnesmoore, Albina Gibadullina, Dustin Gray, Zoe Power, Daniel Gamez, Tanaz Dhanani, Christa Yeung, Sherry Yang, Rachel Brydolf-Horwitz, Sam Walker, Craig E. Jones, Bobby Malone, and Manu Kabahizi.

Thanks also to Ginny Kilander and the other helpful folks at the American Heritage Center for facilitating access to their archival treasures. I am also grateful to Sara Georgi and Abby Freeland at WVU Press for steering the book to production, to Joseph Dahm for beautifully meticulous copyediting, and to Celia Braves for indexing.

And for helping me to work out what my ideas might possibly mean, and for everything else, forever, I am grateful to Indu.

CHAPTER 1

Ackerman's Frontier

Consider three quotes on matters of geography, technology, and communications, separated by more than half a century:

> Urban technology is the new frontier, a space into which big business is entering: after defence, health care, telecommunications and utilities, the management of our cities is being altered by digital and bio-technologies—touted by the World Economic Forum . . . as the fourth industrial revolution.[1]

—————

> I regret to say that my Lowell Institute lectures, "The Art of Coercion," have never been printed, and exist at the moment only in manuscript form. Although the first lecture, "The Theory and Practice of Blackmail," received considerable circulation in an earlier period, it is not now available for distribution. I hope in the near future to remedy this situation and will send you a copy of that lecture, at least, when possible.[2]

—————

> The scientific revolution we have been going through is being accompanied by a revolution of rationalism in our economic structure. Indeed, it has been called a "second Industrial Revolution," with effects already very profound for all humankind. Industrial engineering years ago removed the individual decision making of the artisan.

"Cybernation," or systems design and engineering, are now rapidly moving individuality from "middle management" decision. This development is part of the social problem of automation. Not least, systems design and engineering, through the nation's defense program, is having a dominant role in domestic political affairs and international relations. Research approaches have even been made toward understanding the process of human thought itself.[3]

The first quote comes from an editorial in the fall of 2016 by Richard Shearmur, then editor-in-chief of *Urban Geography*, providing an update on the latest "technophilic" discourses of Google, Cisco, Uber, Airbnb, and the other corporations of contemporary digital disruption. Shearmur exposes the deep contradictions of urban technology lobbyists who demand that governments "get out of the way" in terms of regulation while aggressively harvesting public resources through R&D investments and pressuring governments to pay as clients of utopian "fourth-revolution" big data services and incessant updates. For technology companies, cities are now viewed as "untapped zones of extraction,"[4] while intensified neoliberal marketization and shrinkage of the public realm drives the consolidation of what Andy Merrifield diagnoses as "parasitic capitalism."[5]

The second quote is from a letter written in April 1963 by Daniel Ellsberg, intelligence analyst inside the RAND Corporation defense think tank. A few years after he wrote this short, perfunctory letter—you wonder, how many of those requests did he have to reply to back in those days before email?—Ellsberg famously leaked the Defense Department's massive internal study of the impossibility of winning the Vietnam War. The *New York Times'* publication of the Pentagon Papers changed the U.S. presidency, set in motion key elements of what came to be known as Watergate, and exposed some of the most horrid details of American imperialism. Ellsberg has written a lot over the years, but if you want to see his most recent thoughts, you can visit the website maintained by his son (www.ellsberg.net), or you can follow him on Twitter at @DanielEllsberg.

The third quote comes from one of the key figures of the "newest 'new geography' of the twentieth century,"[6] who has now been almost entirely forgotten: Edward A. Ackerman. This is from a published version of a talk he gave to the Association of American Geographers in September 1963. Do you see the typo in his quote?[7] To be sure, systems engineering and the Greek *kubernētēs*, "steersman" of Norbert Wiener's "theory of messages,"[8] is fundamentally about dynamism, change, motion, and acceleration—and in that sense, cybernetics is indeed about "moving individuality." But that's not what Ackerman had in mind. He meant *removing* individuality.

We could dismiss this typo as nothing more than a trivial error. But Ackerman was a legendary editor, and his friend and commanding officer Edward Ullman was constantly reminding him that attention to detail could win or lose a war: "Good editing, I still think, is about the highest level work in OSS," Ullman once wrote about their work together in the Office of Strategic Services.[9] Thus we can learn something if we pause to consider what this copyediting error means in the context of geographical knowledge production in 1963. One aspect of that context involves the mundane mechanics of the writing process. We know that the typo was a product of (mistaken) human action of some sort—a misreading or a few omitted keystrokes by his typist, followed by a human proofreading error by Ackerman himself. Today we'd also need to consider the more-than-human possibilities of auto-correct, spell check, grammar check, or the speech-recognition algorithms that have become part of the cybernation of human communication practices.

A second aspect of the context of 1963 involves the grand scientific narrative Ackerman was presenting for his audience at the Awards Banquet session of the Fifty-Ninth Annual Meeting of the Association of American Geographers, in Denver, Colorado, on September 4. Ackerman was appointed as honorary president of the association, and his presidential address was a panoramic manifesto for geography's role in the stunning scientific advances and possibilities that were then so clear. His lecture, "Where Is a Research Frontier?," began with a review of the twentieth century's achievements—the revelation of more than a billion galaxies in the universe, the general theory of relativity, the splitting of the atom, space orbit, the discovery of the biochemical processes of heredity, "the developments in engineering made possible by the Manhattan Project"—before considering the more modest contributions of geography. The cybernation passage appears in a section titled "Systems Methods Are Changing Society," where Ackerman advocates General Systems Theory as a means of unifying geography's focus on a central purpose: "nothing less than an understanding of the vast, interacting system comprising all humanity and its natural environment on the surface of the earth."[10] Only a few lines later he speaks of "understanding the process of human thought itself," mentions the scientific advances of "manipulating some aspects of society, like consumer demands, in a more or less controlled fashion," and portrays a variegated yet unmistakable process of "rationalization" across the United States, Europe, Japan, and the Soviet Union.

In this short book, I analyze geography's quantitative revolutions—past and present—through the ideas of Edward Ackerman, especially his wildly ambitious *Frontier* lecture. This purpose, I admit, may seem like an esoteric digression from the urgency of "now" that shapes our present worlds of geographical theory, technology, and practice. In what Geoffrey Martin once sardonically labeled "that now receding genre in American geography, the history of geographic thought,"[11] few geographers today even remember Ackerman, and most outside the field who do remember him are unaware that he was a geographer. Ackerman is even more obscure than his wartime boss, Richard Hartshorne. Hartshorne was once among the most influential and widely read geographers in America, but not too many years later, the endless reissues, revisions, reflections, and remixes of his masterwork *The Nature of Geography* had become scriptural orthodoxy—always dutifully cited but never really read or used in any significant way, what Peter Gould once savaged as "authoritarianism without authority."[12] Today, "Hartshorne" is only a proper noun, a keyword for a historical reference point rather than a theorist to be studied—although the exceptions can be quite revealing: Hartshorne remains part of "The Essence of Geography" in the syllabus for Geog 571, "Intelligence Analysis, Cultural Geography, and Homeland Security," a required course for the Geospatial Intelligence option of Penn State's Online Master of GIS Program.

Ackerman's *Frontier* was just one of more than fifty presidential addresses to one association (self-identified geographers) in one place (the United States) in one era (the twentieth century after the Second World War). His talk was one among the twenty-three agenda-setting presidential talks officially classified as "History and Philosophy of Geography" manifestos.[13] Why, then, should we pay attention to Ackerman's speech? One justification is scholarly and genealogical. Ackerman is formally acknowledged in the literature as "one of the first geographers to point to the rise of systems research throughout the scientific world after World War II."[14] Before that, however, Ackerman played an early, central role in the reconceptualization of a fundamental geographical concept—the region—as the scholarly priorities of academic geography changed with the development of America's military-industrial complex.[15] If we accept the judgments of the science historian George Dyson, who identifies the nuclear computational breakthroughs of 1945 as the birth of a new "universe of self-replicating code" and militaristic cybernetic logics that connects today's everyday practices of Google searches and Twitter feeds with the cognitive-computational worlds envisioned by Alan Turing,[16] then Ackerman merits serious consideration as a counterpart for the field of geography. Ackerman also lived through a maelstrom of professional and personal dramas that shaped the institutional position of the discipline in this crucial period. Revisiting Ackerman's life and work therefore helps us to recover forgotten episodes from our field's histories: it's been quite a few years since Trevor Barnes began a historical analysis by citing Ron Johnston's lament that "the quantitative revolution is rapidly receding from human geography's institutional memory."[17] Our literatures provide an important reflection of our struggles for the "heart and soul" of our discipline, for resources, and for the "hearts and minds" of colleagues, allies, and students.[18]

A second set of justifications, however, is more immediate,

contemporary, and mundane. Our present, taken-for-granted geographical worlds seem to fulfill Ackerman's remarks about "cybernation," the "social problem of automation," and the "research approaches" to "understanding the process of human thought itself." In a world where more than 215 billion emails are sent every day—and two-thirds of users worldwide access their email with a mobile device—why should we waste our time reading the typographical errors of a long-dead geographer?[19] Why are *you* spending *your* time reading *my* reflections on Ackerman's speculations on the "rationalizing" scientific management of consumer desires and public opinion, when you could just swipe a thumb to join more than a billion daily active users who use a mobile device to access Facebook,[20] where your news feed might very well enroll you into the latest algorithmically randomized control trial "61-million person experiment in social influence and political mobilization,"[21] or perhaps an experiment of socially networked "emotional contagion" that can occur even without your direct interaction or awareness?[22] The dominant analytics firm ComScore, with their cheeky corporate motto "Precisely Everywhere," tells us that in one recent month Google Maps had 64.49 million unique visitors by mobile device and that Apple Maps had 42.07 million; about two-thirds of all smartphone users use mapping apps. These figures are only for the United States, reflecting the enduring American hegemony of the world's "data rich" landscapes of mapping and measurement,[23] but don't they provide one kind of estimate of how many people today might be considered "geographers"? And then we smile at the thought of one Canadian delegate's reaction to that "gigantic" meeting in September 1963, when eight hundred geographers gathered over four days for a conference program featuring a hundred eighty paper presentations in the "sumptuous" new Denver Hilton Hotel.[24]

One premise of my argument is that, simply put, the quantitative revolution won. Today we are living in a mobilized, revolutionary, and evolutionary quantification of a cybernetic geography that, in 1963, was just getting under way. The early 1960s are therefore an especially fertile era for us to study to understand our bewildering present. Our current revolution is moving individuality by the billions, and, thanks to all the algorithms, bots, and artificial intelligence advances, it is removing individuality as well. This is why I am obsessively reading and rereading Ackerman's 1963 presidential address. I wasn't there in Denver, and (except for a tiny and diminishing share of the human eyes that may read these words) neither were you. But we are living in a world mutated by the General Systems Theory that Ackerman was advocating. Our lives in today's "Data Revolution"[25] are saturated with the all-too-familiar routines of a cybernetic existence. We line up at the self-check-out kiosks that are proliferating at so many retail stores. We talk to artificial intelligence voices when we call customer service lines. We stare at our smartphones rather than making eye contact with anyone nearby, and we prioritize our every waking moment according to the endless streams of emails, text messages, status updates, and tweets. We wander through Pokémon Go's addictive augmented realities. After we escape the retail self-check-out kiosks, we head to the airport and line up at the self-check-in kiosks for passport and biometric authentication. We are living embodiments of the "social problem of automation" that Ackerman foresaw. As capital and code augment, magnify, and then replace what human geographers do, we confront a paradox of a world where human geography is more popular than ever before—and where it is no longer clear what, if anything, we need human geographers to do.

I believe that human geography is more urgent than ever before, in an era of accelerating transformations of social relations and spatial structures. Human geography matters not because of the essence of a particular body of "external" spatial-scientific realities (more of which are coded and codified in robotic data-processing assemblages), but for the sake of generations past and present of deeply flawed humans struggling to understand our place in the world. This book is not, however, a sustained argument for human geography: I am in awe of dozens—no, *hundreds*— of lengthy, detailed disciplinary visions produced through the generations that are far more compelling than anything I could hope to produce. Instead, what is offered here is a brief strategic intervention within a field caught up in a period of dramatic, accelerating change. I will read Ackerman's *Frontier* to disentangle a set of (largely overlooked) continuities and ruptures of geographical thought over the past century. Today's sudden, explosive popularity of human geography threatens to replicate a particular kind of spatial knowledge implicated in the most dangerous combinations of capital, code, and coercion. My method involves a blend of Foucault's technique of "discursive instaurations" as applied to the historically and geographically situated "author function," Barnes's "geo-historiography" connecting "the history of the discipline's ideas with where they were produced and traveled," and Hartshorne's narration of the "life and times" of a text as a "bibliobiography."[26] I'll outline a story in three parts. First, we'll explore how geography made Ackerman. Second, we'll study how Ackerman made geography—how he tried, and partly succeeded, in remaking the means and ends of geographical knowledge production. Third, we'll consider how his General Systems Theory has evolved into a pervasive yet unstable operating system for a networked society of coercive consumption, automated cybernetic social relations, and militarized spatialities of surveillance. Taken together, I show that the achievement of the 1963 *Frontier* vision entails a machinic, evolutionary erasure of human agency, subjectivity, labor, thought, and understanding.

I should add one more caveat. Daniel Ellsberg's "Art of Coercion" lectures are, in fact, easy to find on the internet. But the cursory letter I've quoted is the only Ellsberg correspondence I could find in the Ackerman archives. Sadly, we don't find any evidence of direct connections to Ackerman's *Frontier*. But my use of Ellsberg's words is more than clickbait. Ellsberg was working in the same evolving informational infrastructures that shaped Ackerman's thought, and we have the good fortune that Ellsberg has lived much longer than Ackerman—and has continued to excavate the lost and suppressed histories of America's military-industrial complex. In his most recent book, Ellsberg reveals that at the same time Ackerman was in Denver and D.C. advocating for an omniscient, rationalist cybernetic geography, Ellsberg had been traveling throughout the Pacific, visiting the airbases where nuclear-alert bombers were stationed, ready to receive launch codes. Meeting with base commanders and pilots, Ellsberg learned about a series of fatal dangers in the control systems for the massive attack fleet. Launch authority was widely delegated across geographical territories where communications outages would, under certain crisis conditions, almost certainly be perceived as evidence of a Soviet strike. There were no provisions for recalling planes launched in error, and indeed all safeguards were designed to prevent any reversal of an "Execute" order for nuclear attack. And all these decision points were implemented in an entrenched military culture encouraging bold, decisive action even in defiance of formal protocol. Ellsberg spoke with a major at a base in Kunsan, South Korea, with near-daily communications outages, and they discussed a scenario where international tensions or a report of a nuclear explosion somewhere in the western Pacific could lead a base commander to decide (in violation of explicit directives) to initiate a precautionary, first-stage launch of at least some of the bombers under his command. Ellsberg asked him what would happen next. The major replied,

> Well, you know what the orders are. They go to a rendezvous area and fly around, waiting for further orders. They can do that for about an hour and still have enough fuel to get to their targets or to come back. If they don't get an Execute message, they're supposed to come back. Those are their orders.[27]

But at the rendezvous area, the planes would be out of communication with the base, the commander had earlier explained. Ellsberg asked what would happen next:

> The major said, "If they didn't get any Execute message? Oh, I think they'd come back." Pause. "Most of them."
> The last three words didn't register with me right away because before they were out of his mouth, my head was exploding. I kept my face blank but a voice inside was screaming, "Think? You *think* they'd come back?!"

This was their commander, I thought, the one who gave them their orders, the man in charge of their training and discipline. As I reeled internally from that response, the next words, "most of them," got through to me.

He added, "Of course, if one of them were to break out of that circle and go for his target, I think the rest would follow." He paused again; then he added reflectively, "And they might as well. If one goes, they might as well all go. I tell them not to do it though."

I managed to keep a straight face.[28]

Back in the United States, one day Ellsberg and his boss at the Pentagon went into the District to see a new movie "for professional reasons": *Dr. Strangelove*, the hilarious yet terrifying portrayal of what happens when a paranoid base commander, Jack D. Ripper, issues the authenticated codes to send pilots off to attack the Soviet Union—and then the president learns that the attack can't be reversed. The entire system of attack authentication codes was designed to ensure that a weak, indecisive commander-in-chief could never have second thoughts and forsake a first-strike advantage. Ellsberg's travels through the networks of bases across the Pacific, and his top-tier security clearance for access to classified information in Washington, had allowed him to see the horrific anatomy of a literal "Doomsday Machine," optimized for planetary "omnicide"—and here that precise scenario was on the screen, in Stanley Kubrick's dark-humor masterpiece, featuring George C. Scott, Slim Pickens, James Earl Jones, and (in several roles) Peter Sellers. "We came out into the afternoon sunlight," Ellsberg reflects, "dazed by the light and the film, both agreeing that what we had just seen was, essentially, a documentary. . . . How, I wondered, had the filmmakers picked up on such an esoteric, highly secret (and totally incredible) detail as the lack of a stop code, and the alleged reason for it?"[29] Ellsberg later learned that one of the screenwriters, the author of the novel on which the film was based, had served as a flight lieutenant and navigator with Britain's Royal Air Force Bomber Command: "That suggests Bomber Command's control system had the same peculiar characteristics" as the U.S. Strategic Air Command, Ellsberg reflects, "and probably for the same underlying reasons."[30]

Ackerman's commanding officer Ed Ullman would have been impressed with the good editing, but for Peter Bryant George (1925–1966), the novel and screenplay involved a bibliobiography poised between life and death. George had been deadly serious when he wrote *Red Alert* in 1958 under the pen name of Peter Bryant—released in the United Kingdom as *Two Hours to Doom*; he initially hated Kubrick's plan to render the story as satire for an American audience that had already been numbed by too many straight-serious apocalyptic nuclear thrillers. A few years after the film's release, deeply pessimistic about the future of an atomic world as he worked on a new novel, *Nuclear Survivors*, George sat in a chair, put a double-barreled shotgun between his knees, leaned down, and pulled the trigger.

Ackerman's demise was not quite so horrific, and he was never "as close to the button" in the military as George or Ellsberg or even Hartshorne. And yet we have much to learn by exploring archival materials written by those enmeshed in the daily rhythms and worldviews of Cold War American science who are no longer in this world. In Trevor Barnes's beautiful rendition, sifting through such archives becomes "taking the pulse of the dead . . . even though there was no heartbeat, life throbbed on the page."[31] Reading through Ackerman's archives allows us to think through what has changed (and what has not) in everyday infrastructures of information—and how, in today's Trumpian Twitter versions of Watergate and alt-right conspiracy theorists' viral videos on the latest plots of the "Deep State," history is quickly reconfiguring itself through evolving, unstable cybernetic general systems.

CHAPTER 2

The Ackerman Sample

Stories are always told by people, about people, for people.
Geography's story is no exception.

—David N. Livingstone,
"Should the History of Geography Be X-Rated?"

Modern science is built on sampling theory—procedures that allow us to study a carefully defined subset of phenomena and then draw wider inferences about comparable phenomena, to build generalizable knowledge. Generalization became a crucial goal as geography pursued the disciplinary status given to science in the 1950s and 1960s, as geographers sought to produce scientific laws—generalizations that are "empirically universally true," and integral parts of "a theoretical system in which we have supreme confidence."[1] During this period, however, the spatial science that geographers developed relied heavily on a narrow strain of the philosophy of positivism—emphasizing objective, neutral observation, abstraction, and measurement—as well as an ontology that was deeply influenced by the philosophy of Immanuel Kant. Ontology is concerned with the nature of being in the world and for Kant (and then also for Richard Hartshorne as well as the first generation of quantitative revolutionaries) the fundamental essence of being involved space and time. The result was a particular kind of understanding of space that shaped how geography's quantifiers engaged with the statistical methods developed in the physical sciences. The new spatial scientists emphasized that sampling is always an inherently contextual process of abstracting something not just from what the statisticians call a population or a universe, but from a place-based intersection of space and time. When applied to the human realm, these place-based, time-space junctures trace out the life paths—the "lives lived" that eventually become the "lives told" of biography, autobiography, and disciplinary history[2]—that shape the interrelations among individuals, theories, institutions, and sociohistorical epochs.[3] The classical sampling strictures of random selection, then, must give way to a humanistic narrative of selection that allows us to draw contextual inferences, to generalize without neglecting contingency, and to reconcile objective and subjective interpretations of significance that will always be partial, situated, and tentative. Interpretations will also be contested and misunderstood: human geography's most famous breakthrough in spatiotemporal sampling came from Torsten Hägerstrand's "dream" of creating a form of musical notation as an attempt to capture what he thought of as the music, poetry, and drama of thousands of human lives lived through a turbulent century in a rural Swedish parish. But Anne Buttimer saw Hägerstrand's geometric drawings of the three-dimensional paths of individuals' daily lives (two dimensions of spatial location in a landscape, a third dimension tracing a path through time) as a "dance macabre," as a "chilly recording by a detached observer," and even as a "hollow rattle of bones" that offered "no communication of the vibrating sound of full orchestra."[4] Yet biographical approaches to disciplinary history are worth the risks of misinterpretation so long as we are willing to listen and learn from one another. Hägerstrand was horrified when he understood Buttimer's reactions to his methods, and over the years they talked about quantitative and qualitative methods and about objective and subjective interpretations, and they learned from one another. This is just one sample of how the theory and practice of human geography evolves *within* and *across* distinct generations of human geographers. The substance of any purportedly geographical method, explanation, or application is really only a means to an end: what matters is that conversations take place among those who situate themselves as geographers.

The life path that brought Ackerman to the banquet hall of the Denver Hilton in September 1963 was truly enigmatic, an $n = 1$ sample that nevertheless helps us understand the astonishing global epoch of science, war, and modernism at the high point of what Walter Lippman had declared the American century. If Hartshorne came to be known (and then vilified) for his theoretical disciplinary pronouncements,[5] Ackerman was often too busy *doing geography* to devote much time to the question of how geography should be defined. The *Frontier* lecture was a rare moment when he allowed himself to advance a bold programmatic statement.

FROM COEUR D'ALENE TO HARVARD

Edward Augustus Ackerman was born on December 5, 1911, in Post Falls, Idaho, to first-generation Swedish immigrants. His childhood was shaped by nature, study, and personal catastrophe. At age ten, Edward got pneumonia and nearly died, but as he recovered his mother caught the disease from him. She didn't survive. A bit more than a year later, his father, working on the railroad, was electrocuted by a high-tension wire blown down in a windstorm. The young boy, who "then knew what it meant to be all alone in the world,"[6] developed a stoic independence, focus, and drive. He was deeply influenced by the changing geographies of his rural and small-town environment—"He had a chance to see the watershed of the Clearwater River and to enjoy its stands of timber before they were cut, a memory that later helped to fire his interest in resources"[7]—but he also developed a laser-sharp focus on books and schoolwork at Coeur d'Alene High School. In 1930 he won a scholarship to Harvard, initially pursuing English literature. In a first-year course, however, his potential impressed Harold Kemp, instructor in the Department of Geology and Geography, and Ackerman soon left the world of literature to pursue a geography that was poised between historical tradition and modernistic futurism, between geophysical/morphological and human/political approaches. "With an undergraduate record of no grade but A," Kemp would later write in nominating Ackerman for membership in the Association of American Geographers, "he was the high man in each year of his undergraduate residence," winning a series of prestigious scholarships and awards. "After his honors examination in his senior year," Kemp wrote, "the geographers left the voting to the geologists who had flocked to take part in this particular examination, and it was the geologists who thereupon voted to break the rule which had been in force in the division since 1912—a rule against awarding the Summa Cum Laude."[8]

But Kemp was only an untenured instructor, hired the year Ackerman came to Harvard. More important for the future of geography was the way Ackerman impressed Kemp's closest colleague and friend, Derwent Whittlesey, who had come to Harvard from an associate professor position at Chicago in 1928. Whittlesey, who had made significant personal sacrifices in order to accept an invitation from Harvard to build a new program in human geography,[9] quickly became Ackerman's advisor, mentor, advocate, and friend. "His mental capacity is extraordinary," Whittlesey once exclaimed in a recommendation for Ackerman, who continued in graduate work. Ackerman traveled in Europe and North Africa on a scholarship in 1934–1935; did fieldwork in Europe in 1933, 1936, and 1937; and completed his preliminary doctoral examinations in June 1938: the subject areas were regional geography, systematic geography, physiography, and meteorology and climatology.[10] Ackerman completed a doctoral thesis on "Regional Factors" in the fisheries industries of New England with a successful defense in the spring of 1939, despite a bit of hostile questioning by an examiner from the Bureau of Fisheries who was "by conviction a high protectionist," as Whittlesey wrote in support of Ackerman, "whereas every geographer is a free trader."[11]

HARVARD TO HARTSHORNE

Ackerman became Harvard's first PhD earned specifically in *human* geography. He thereby achieved an important milestone of Whittlesey's long-term project to create an independent Department of Geography within the Division of Geological Sciences. Whittlesey nominated Ackerman for an instructor position at Harvard, but he also recommended the brilliant new PhD to Richard Hartshorne, who in September 1941 was put in charge of the Geographical Division of the Coordinator of Information (COI) in Washington, D.C. Whittlesey's recommendation came at just the right time, as Hartshorne had jumped into "the premier decade for

methodological controversy" among academic geographers
at the same time he became involved in America's expanding
military and intelligence infrastructure.[12] It had been
Whittlesey, after all, who had given encouragement after
a 1937 lunchtime meeting with Hartshorne, who angrily
denounced a methodological paper recently published in
the *Annals of the Association of American Geographers*
by suggesting that it should be "consigned to the circular
file."[13] Why not prepare your remarks in a form that can be
published?, asked the diplomatically persuasive Whittlesey,
the editor of the *Annals* who would later be called "one of
the most charming men in the world."[14] Over the next
eighteen months, Hartshorne wrote while Whittlesey offered
commentary, inspiration, and provocation; in August 1938
Hartshorne departed for sabbatical travel in Europe on a
grant from the Social Science Research Fund of the University
of Minnesota. The initial research plans to study border issues
along the Danube collided with real-world events: driving past
Hungarian troops on their way to the Czechoslovakia border
and encountering German forces on the way to the Austrian
frontier to "protect" Slovakia, Hartshorne retreated to the
library at the University of Vienna. He wrote and revised
as "many of the gleanings from the printed word" were
"verified and amplified in discussions between the author
and German authorities"; Hartshorne's immersion in the
"atmosphere of the native heath of geographic methodology"
prompted a "complete revision."[15] By May 1939 Hartshorne
had produced a six-hundred-page manuscript, *The Nature of
Geography*. Whittlesey secured a speedy peer review (from
a single referee, J. Russell Whitaker, another Chicago PhD
who subsequently worked in military intelligence and served
on the National Research Council),[16] accepted the piece, and

took the unprecedented step of publishing it as the last two full issues of the 1939 volume of the *Annals*.[17]

Hartshorne thus had very good reasons to trust Whittlesey's judgment of a rising-star geographer, writer, and editor, and so in September 1941, Ackerman became the first geographer hired by Hartshorne at COI. Ackerman was appointed as chief of the Geographical Reports Section.[18] Thus began nearly a decade in which Ackerman became a major influence in the applied geography of the U.S. military while seeking a permanent academic position in the new Harvard Department of Geography that Whittlesey had been trying to establish. For Ackerman, the first half of the 1940s was divided between teaching in Cambridge and research and writing at COI—which in 1942 was reorganized and renamed the Office of Strategic Services (OSS)—and then, after the war, serving as an advisor to General Douglas MacArthur's occupation of Japan. At COI/OSS, Ackerman deployed his meticulous scholarship and a painstaking attention to detail—a colonel up the chain of command once saw his editorial corrections on a draft report and declared, "Get that man for our editor-in-chief, since he so obviously demonstrated his ability"[19]— to shape the "Geographical Reports" that became the Joint Army Navy Intelligence Surveys (JANIS), later the *CIA World Factbooks*. He returned to Cambridge in 1943 after a leave of absence from Harvard and was promoted from instructor to assistant professor. But his job had become anything but academic: he had been brought back to teach in Harvard's School of Overseas Administration, part of a national network of programs at ten universities designed to prepare officers to administer military government in occupied territories.[20] Ackerman's assignment was to teach the geography of Japan— to officers in classes that sometimes numbered almost four hundred. Ackerman later reflected that he was a "mere beginner" on the subject: "Although I have learned much, I cannot see that it has contributed greatly to what I hope will be my future interest."[21]

MacArthur and Modernization

One part of that future interest involved plans for a book on resources and conservation, and another part was a "paper on geographic methodology" reflecting Ackerman's dissatisfaction with the way Hartshorne "left our subject hanging in his otherwise memorable and impressive work."[22] Ackerman quickly completed the methodology paper, but the natural resources book (eventually coauthored with Whitaker) took several years because the second half of the 1940s prolonged the applied military assignments. Within days of Japan's surrender Harvard's School of Overseas Administration was disbanded, and all the officers received orders to head to the Pacific Coast within a week. "You can imagine the scurrying around which took place," Ackerman wrote to a lieutenant colonel in the Pacific theater, lamenting that the School's Germanic Museum and Semitic Museum had been unceremoniously stripped bare of all books and maps.[23] In the spring of 1946 Ackerman accepted an invitation to serve as an advisor to the Natural Resources Section of General MacArthur's U.S. Supreme Command in Japan. "In spite of all the psychological handicaps that history has imposed on them," Ackerman later wrote to provost Paul Buck at Harvard, "I think that the Japanese have the capacity for understanding democracy."[24] Ackerman's section was charged with assessing agriculture, forestry and forest products, fisheries, mining, and all associated institutions—with an eye toward "improving long-term resource use in Japan."[25] Economic modernization was essential if the Japanese were to develop democratic sensibilities. "It remains for the Allies to help them disprove Goering's perceiving statement at Nuremburg, which they still remember, 'Only the rich can afford democracy,'" Ackerman told Buck; "But it is evident that the assistance must be continued a long time if it is to be effective. For that reason almost every responsible person here wishes for a long occupation."[26]

By this point, the shape of Ackerman's impressive career trajectory seemed well aligned with the expansion of postwar geography. Ackerman had served with distinction at an influential government agency with one of the nation's most prominent geographers, at a time when the OSS employed more than a hundred geographers within a larger Washington, D.C., cohort—at that time "the largest collection of American geographers ever drawn into a common enterprise with other scientists"[27]—that developed a "coherent paradigm" sustained by both close physical proximity in Washington and the strengthening global networks of American empire.[28] In the Natural Resources Section of MacArthur's occupation, Ackerman coordinated all research activities of a staff of 253, and as described by his chief, Lieutenant Colonel Hubert Schenck, "Our decisions directly affect the activities of several thousand Japanese officials and research workers, and they affect the amounts of food, fuels, clothing, and other materials available to the 75 million Japanese."[29] For Ackerman, this applied geographical activity made for "very pleasant circumstances." "At my own request I have a minimum of administrative detail, the usual curse of government work," he told Buck; "There are about a hundred Allied specialists and about a thousand Japanese whose services I can call on, so I have no lack of research assistance. My largest problem is that of finding time to make full use of the facilities I have."[30]

"Geography Is Not a University Subject"

But the scale of academic geography was another matter. Ackerman desperately wanted a permanent, tenured position at Harvard; he also wanted to help achieve Whittlesey's dream of an independent Department of Geography at the pinnacle of America's university hierarchy. Neither came to pass. The attempt to promote Ackerman to associate professor intensified a series of institutional struggles within Harvard's Division of Geological Sciences as well as high-stakes disciplinary conflicts in an external review committee and at the highest levels of the university. Everything culminated in a decision in February 1948 to "taper off" all commitments to geography. It was, the legendary French geographer Jean Gottman later reflected, "a terrible blow," from which U.S. geography "never completely recovered."[31] Harvard president James Conant's famously condescending defense of his decision in response to critics—"geography is not a university subject"—echoed throughout the academic world, undermining the credibility of the discipline for decades. In the wake of Harvard's decision not to create a geography department, programs at Yale, Stanford, Northwestern, Columbia, Chicago, Michigan, Pittsburgh, and Temple were eliminated.

Harvard's decision pushed Ackerman ever deeper into the circuits of advising, consulting, and policy formulation. In August 1948 he accepted a faculty position at Chicago, to begin in January 1949. The interim period was spent on a second tour of duty as a visiting expert consultant to the Allied Supreme Command in Japan, where "particularly significant" among his personal conferences were those with General MacArthur and Emperor Hirohito—"The significance of these conferences is that your work is fully appreciated at the highest levels," Schenck assured Ackerman.[32] At Chicago, he was "at the University so little that a student had to be alert to take advantage" of Ackerman's brilliance.[33] His advisory and consulting portfolio grew rapidly: he undertook a review of resource management methods in the Missouri River Basin for the Hoover Commission, served as chief geographer for the President's Water Resources Policy Commission, headed a natural resources and public works branch of the Bureau of the Budget, served as assistant general manager of the Tennessee Valley Authority, and headed the water resources program of the think tank Resources for the Future. In 1958 he accepted a position as assistant executive officer at the Carnegie Institution of Washington, and two years later he became executive officer, a strategically influential position from which he would guide geographical contributions to public policy for the rest of his life. As Gilbert White later noted, Ackerman "probably was the first geographer to engage himself very widely in the new breed of private research groups that became such a large part of the American research establishment during the 1960s,"[34] and it was thus from this high point that Ackerman accepted the honorary presidency of the Association of American Geographers.

Carnegie and Scientific Frontiers

The *Frontier* lecture thus came from a distinctive perspective, as Ackerman had succeeded the "superbrain" Vannevar Bush, the Carnegie president who had directed the weapons-innovation hotbed Office of Scientific Research and Development (OSRD) during the war.[35] From distinguished service in the OSS and briefings of General MacArthur to frustrated ambitions at Harvard and executive authority at Carnegie, Ackerman was a living embodiment of the "revolution in the conduct of research" described by another general: Dwight D. Eisenhower. Ike's January 1961 farewell address is universally remembered for his concise label for a new mode of politics, science, and technology—the military-industrial complex—and there is no doubt that Carnegie was one of its key nodes;[36] Eisenhower's speech was also a premonition of alternative meanings in White's praise for a geographer's pioneering role in the "new breed" of private research groups. Eisenhower warned of the dangers that a massive federal defense establishment posed to precisely the kind of scholarly path that Ackerman had once tried so hard to pursue at Harvard: "The free university, historically the fountainhead of free ideas and scientific discovery, has experienced a revolution in the conduct of research. Partly because of the huge costs involved, a government contract becomes, virtually, a substitute for intellectual curiosity. For every old blackboard there are now hundreds of new electronic computers. The prospect of domination of the nation's scholars by Federal employment, project allocations, and the power of money is ever present—and is gravely to be regarded."[37]

At this point there's a fascinating paradox. In his presidential address to the AAG, Ackerman mentions none of these historical circumstances—which pervaded elite and professional consciousness amid the generational transition from Eisenhower to Kennedy's "New Frontier." Instead, Ackerman offers a polished, sterile exemplar of the "new geography" that was emulating the ahistorical discourses that were being consolidated in science policy in postwar America.

Those discourses, refined through the expanding review infrastructures of grant and contract funding pools offered by the federal government and private research foundations, emphasized the technological achievements and possibilities of scientific innovation—while downplaying or entirely ignoring the social contexts and biographies of scientists *as individual human beings*. In the new wave, grant proposals and published research were to have fewer names and more equations: appeals to the authority of mathematics began to supplant appeals to the authority of words written by a field's recognized philosophers; at the extreme, for the purest of sciences, the fewer words the better.

Through the 1950s, America's sudden ascension to global imperial leadership was sustained by a proudly modernist amnesia (an excess of historical awareness was always reigniting Europe's endlessly simmering cycles of war) joined with an anticommunist paranoia of free ideas that could all too easily lead to sedition. One result was an obsession with objectivity and neutrality to guard against accusations of politicized research. Another was the rapid embrace of positivist philosophies of knowledge and quantitative methodologies by the social sciences at the precise moment when physical scientists were having second thoughts. Quantum physics destabilized the neat separation between observer and observed, such that quantification, observation, and positivist logics could become unhinged. Even the most strident defenders of positivism, the philosophers of the Vienna Circle in the 1920s and 1930s, found it increasingly difficult to provide any rational justification for the mechanical logics of positivist quantification that were coming to dominate the mundane routines of what Thomas Kuhn critiqued as "normal science."[38]

By the time Ackerman came to the lecture hall in Denver in September 1963, he had been living through these transformations in the culture of science for two decades, at the interface between an academic discipline and a Washington, D.C., technoscience complex skeptical of the relevance or even the existence of a coherent field of geography. Indeed, in the months he spent preparing *Frontier*, Ackerman was juggling his busy schedule of Carnegie responsibilities with another key role—chairing the Ad Hoc Committee on Geography of the National Academy of Sciences' National Research Council—that was yet another attempt to promote the merits of geography to the disciplines and institutions with the *real* power and authority. This effort was dominated by a vision of an impersonal, objective, and universal science—a hard-fought but inevitable achievement of rationality and progress culminating in accelerating technological advances that seemed to be on the edge of a new possibility, the refined automation of the enterprise of science itself. While the six-member NAS/NRC Committee included some diversity of views, it was clear that Ackerman's sympathies lay with the scientific ambitions of two young, bold members: Brian J. L. Berry (then at the University of Chicago) and Edward J. Taaffe (Ohio State University). In a draft working paper submitted to Ackerman for the committee's discussions, Berry and Taaffe offered what might best be understood as a theory of automated positivism: "We now stand on the verge of developing man-machine systems such that concepts can immediately be referred to a computer system and a data bank for testing and results."[39]

Herein lies the paradox of *Frontier*: a man who had been in the maelstrom of so many personal conversations with so many professional human geographers at Harvard, in

Washington, and in all the far-flung nodes and networks of geographical thought and action in a world at war presents a lecture that makes little mention of the individuals involved in the struggles to define the purposes and ideas of the field. Ackerman dutifully traces a brief, selective genealogy of Carl Sauer and the California School of cultural geography, Harlan Barrows's plea for geography as human ecology, and Hartshorne's use of Humboldt, Ritter, and Hettner to defend a definition of geography as areal differentiation—the "accurate, orderly, and rational description of the variable character of the earth surface."[40] But Ackerman distills the history into few words, with a minimum of contextual, biographical, or philosophical narrative. Then he pivots to his central claim, that geography must become a science focused on an "understanding of the vast, interacting system comprising all humanity and its natural environment on the surface of the earth." This requires deepening the field's commitment to mathematical statistics, choosing research problems according to "the advancing frontier" of the behavioral sciences, and using General Systems Theory both to unify the physical and human sides of the field *and* as a litmus test for prioritizing fields to engage in interdisciplinary inquiry.[41] The really important advances in problem solving in the history of science are few and far between, Ackerman emphasizes, citing writing, Arabic numerals, analytical geometry, calculus, and "the combination of techniques that comprise systems analysis." "There was a time, perhaps just after the Second World War, when the inclusion of systems analysis in such a list might have been controversial," he observes; "That is no longer true. Systems, as you know, are among the most pervasive and characteristic phenomena in nature." For Ackerman, anything and everything—including everything about each individual human being—could be subdivided into its constituent elements and hierarchical relations to be modeled with techniques applicable to both natural and human phenomena: "Each human being, man or woman, is a system, that is, a dynamic structure of interacting, interdependent parts. Perhaps that is less appealing than a poet's definition of a pretty girl, but it has meaning in that it relates the girl as a system to all other systems, such as a colony of ants, or a city, or a business corporation."[42]

Ackerman later returns to the misogynist metaphor of a gendered vivisection to illustrate the importance of hierarchically nested scales of systems: "The pretty girl, if you like, can be broken down into an astonishing number of subsystems like any complex being."[43] Yet Ackerman makes no mention of the biologist who developed the theory (von Bertalanffy), instead citing a short review article from the journal *Management Science*.[44] Likewise, Ackerman identifies "cybernation" as an industrial revolution on the verge of understanding and thus engineering human thought itself—conveying a meaning ambiguously positioned between a "nation of cyborgs" and the formal definition of "cybernetics." Moreover, he does this without ever mentioning his fellow Harvard alum, the mathematician Norbert Wiener, who had fully developed the theory and discourse of cybernetics.[45] Ackerman was thus led astray by the widespread yet misguided popularized version of cybernetics: by the time Wiener received the National Medal of Science in 1963, he was almost universally misunderstood as advocating the automation of human life, and few seemed to notice his clear warnings about dehumanization and the interaction of entropy, unpredictability, and irrationality in the "contingent universe" revealed by revolutionary post-Newtonian quantum physics.[46] Ackerman's cybernetic systems thinking was the safely noncontingent universe—linear, progressive, rational, scientific, and evolutionary—that had refined, simplified, and purified the chaos of the recent history of geographical thought into the new, modern geographical synthesis. I suspect this had a lot to do with the deep sense of insecurity born of the discipline's marginal position in a fast-changing academic division of labor as well as the fact that regional geography had quickly been equated with casual, subjective description.

Ackerman gives no hint of the enormous philosophical battles that geographers had been fighting over the meanings and significance of areal differentiation. These battles were intensely passionate and personal, even when filtered through the staid scholarly discourse of academic journals.[47] Ackerman ignores it all and uses a brutally efficient one-liner—"We no longer debate about whether geography can construct 'laws'"[48]—to settle the fights that had been raging for a decade on the dichotomy between "regional/idiographic" versus "systematic/nomothetic" paths to geographical knowledge. And Ackerman makes no mention whatsoever of the most important influence of *The Nature of Geography*: Hartshorne's textually explicit and genealogically forensic appeal to the individual human scholarly authority and philosophical prestige of Immanuel Kant.

This is a remarkable omission. Our disciplinary lore is that Hartshorne built his theoretical perspective on Kant, but this claim is questionable: Hartshorne's deepest theoretical influences came from Alfred Hettner and other German geographers in the century after Kant's death. Hartshorne's *Nature* was not, in fact, a full engagement with Kant's philosophy of space, nor with Kant's lectures on physical

geography. But Hartshorne repeatedly cited Kant to claim
prestige for geography, and this is what made *Nature* famous
among the first generation of American geographers trained
as geographers who struggled to secure their position in
the academy. Hartshorne's collection of textual fragments
of Kant's conceptualization of geography and history as
"exceptional" disciplines defined by the axiomatically
universal dimensions of space and time became the primary
weapon in attempts to defend the discipline's academic turf
at Harvard and beyond in the 1940s. Ackerman knew this
well. He had lived through the failures of the appeal to the
philosophical heritage of the oracle of Königsberg and the
widespread neo-Kantian revival of the early twentieth century.
Indeed, although nobody fully understood the implications
at the time, twenty years earlier Ackerman had managed
to theorize the architecture for a kind of modernist, post-
Kantian idealism on a planetary scale—a mode of thinking
and knowledge production that blends a twisted form of
Kantian metaphysical and epistemological strains of idealism
with the cybernetic essence of what we would today recognize
as big data, social media, crowdsourcing, and cloud-based
collaborative investigation. That was Ackerman's "paper on
geographic methodology," written as a way of making sense
of his COI/OSS applied geographical work in light of his
frustrations with how Hartshorne had left the ontological
nature of academic geography "hanging." Published in the
December 1945 issue of the *Annals* as "Geographic Training,
Wartime Research, and Immediate Professional Objectives,"
that paper was widely discussed for its hectoring tone and its
pragmatic admonishments: American geographers suffered
from an "almost universal ignorance of foreign languages,"
a "bibliographic ineptness," and a "general lack of systematic

specialties"; if "our literature is to be composed of anything more than a series of pleasant cultural essays," Ackerman warned, we had better learn to specialize, and do it fast.[49] But Ackerman's extension of Hartshorne's neo-Kantian thought was completely overlooked. It was easy to miss thanks to the way Ackerman's writing was being shaped by the depersonalized, ahistorical, and modernist discourse that was rapidly evolving through the communicative networks of the military-academic-industrial complex. The massive exegesis of the writing, thought, and influence of historically prominent individual thinkers in a discipline—what made Hartshorne's *Nature* so influential in 1939—became obsolete after the war. Although Hartshorne's massive tome remained influential for decades, few read it, fewer read it carefully, and—especially in the postwar academy in the United States—fewer still had the time or inclination to actually read the works of the individual scholars over the centuries that Hartshorne had cited.[50] That was the old, European past. Postwar America was about science, efficiency, and the future. And so Ackerman's 1945 planetary Kantianism was developed not by meticulous citational genealogies or interpretive discussions of the words and meanings intended by Kant, Hettner, or others in the vast pantheon of neo-Kantians. Ackerman's epistemology was unnamed, a new modern synthesis of an objective, inevitable, and rational science transcending the limits of individual human thought—just like the 1963 *Frontier*.

I am fully aware that the mention of Kant at this point is a serious provocation. Successive generations of neo-Kantian revivals have reproduced commitments to a certain kind of idealism as "a condition of the Western mind" for more than two centuries, and varied, often contradictory species of Kantianism have been credited as inspiration for all sorts of positivism, humanism, and Marxism that have shaped modernity.[51] Even today, Kant is invoked as a point of reference for "all who seek to know the world and its earthly contents,"[52] at every possible spatial scale. At one extreme, the neo-Platonist Ralph Cudworth's influence on Kant reproduced a brain-level lineage of "cognoscitive powers" and a doctrine of "referentialism" (correlating "inward ideas" to cognition of external events) that still constrains evolutionary theories of language.[53] At another scale, the "new liberal imperialists" of post–Cold War and post-9/11 America sought to legitimate military intervention through Kant's philosophy—a "repackaging of longstanding U.S. aspirations toward empire as neo-Kantian, 'liberal-democratic' world peace."[54] Bush's Project for a New American Century—"Kantians with cruise missiles"—morphed into philosophical justifications for Donald Trump in the new journal *American Greatness*, now equating the "globalist world order" with a totalitarian progressivism that threatens to "subordinate America to an ever-expanding collection of international institutions" rooted in a Kantian counter-Enlightenment.[55] In opposition to the right-wing hijacking of Kant, Joel Wainwright has worked to bring a postcolonial Kant to bear on the militant empiricism of the Bowman Expeditions, a venture launched in 2004 by the president of the American Geographical Society to combine participatory fieldwork with cutting-edge geographical information science representations of indigenous territories of Oaxaca, Mexico (with funding from the Foreign Military Study Office of the U.S. Army).[56] This is a theoretical struggle of enormous political significance (at the scale of individual minds and a planet of billions of thinking beings), and thus I wish to proceed carefully with my assertions of a latent Kantianism in a figure involved with the early stages of geography's quantitative revolution and the military-industrial complex. Thus to understand the way Ackerman extended neo-Kantianism—which in its transcendental pragmatist North American form maintains a faith in the capacity of reason to build a whole from its parts, and which holds that "to understand Kant is to go beyond him"[57]—we need to work backward a bit, to explore the bibliobiographical contradictions behind the thought expressed in his *Frontier* modern synthesis.

CHAPTER 3

Contradictions of "Mental Structuring"

Not long ago, the philosopher of mind Mark Rowlands suggested that if we accept a broad definition of neo-Kantianism as "the view that there are activities of the mind whose function is to structure the world," then it is "difficult to imagine anyone who is not a neo-Kantian."[1] Rowlands's claim helps us appreciate our inherited memories of geography's quantitative revolution as a sort of Big Bang of instant innovation. Ackerman straddled a transitional zeitgeist of scientific discourse in the 1950s and 1960s, when awe for the personal achievements of heroic individual *scientists* was held in dynamic tension with the impersonal nature of *science* as a collective, intergenerational enterprise—in Max Planck's famous formulation, "science advances one funeral at a time." Ackerman's eloquent *Frontier* lecture exhorted geographers to commit to the scientific method and to pursue strategically chosen "cross-disciplinary communication," connecting only with those fields that would best prepare a long-isolated geography to join the "unity of scientific effort" that had produced the "panorama of glorious scientific achievement" witnessed in the physical sciences.[2]

There was a risk here: as geographers devote greater time and effort in pursuit of the unity achieved in the more powerful physical scientists, wouldn't the intellectual exodus render geography an even more diffuse, incoherent enterprise, with no justification for a separate status or identity? Ackerman's solution, relying on Hartshorne, seems Kantian: geographers *think differently.* This is a distortion of Kant's true position, which justified a discipline of geography because space is an a priori intuition of the capacity of human reason. Yet Ackerman proceeds to make a case for geography's position in "the intuitive side of science," citing a speech by the mathematician Warren Weaver, who had recently retired from long service as director for natural sciences at the Rockefeller Foundation: "Science is, at its core, a creative activity of the human mind which depends on luck, hunch, insight, intuition, imagination, taste, and faith, just as do all the pursuits of the poet, musician, painter, essayist, or philosopher."[3] "But there is more to it than this," Ackerman tells the audience.

> The mind of the scientist, no less than that of the poet or musician, must be structured by thought and experience before it reaches the creative stage. Some persons are able so to structure their minds more easily than others. It has been said, for example, that Irving Langmuir always saw matter, of whatever form, wherever he was, in terms of its molecular structure, thus opening the way automatically for his many remarkable insights. Every scientist does this in some degree. *There is no doubt that there is such a thing as "thinking geographically." To structure his mind in terms of spatial distributions and their correlations is a most important tool for anyone following our discipline.* The more the better. *If there is any really meaningful distinction among scientists, it is in this mental structuring.* It is one reason why we should approach the imposition of analogues from other fields, as from physics,

with the utmost care. The mental substrate for inspiration does differ from field to field.[4]

There are some dubious claims here, most likely because Ackerman's view of the dichotomy between systematic and regional geography was distorted by Hartshorne. In Hartshorne's reading of Kant—or at least his early read of the late nineteenth-century neo-Kantians—a unity of human knowledge was achieved through "Kant's idea of *raum* (area or space), a holistic notion in which the findings of subjects ranging from anatomy to zoology could be held together as parts of a single whole."[5] Hartshorne had parlayed that idea into narrow, defensive, and backward-looking conceptualizations of the region and of regional differentiation. These notions were insufficient to fulfill the expectations of applied military service or competing scientific disciplines, so Ackerman stressed "systematic" geography. And despite the need for "utmost care," Ackerman eagerly embraced the imposition of analogues from the physical sciences. Geographers must join the wider scientific quest for the formulation of universal scientific laws, and although in interdisciplinary work "we do retain an identity by structuring our minds to handle spatial distribution patterns in all their complexity,"[6] that identity is put into service for General Systems Theory—yielding a unified science of objectivity, inevitability, and rationality.

Each of these goals, however, betrays intimate contradictions in the human life path and structuring of mind in the human geography that produced Ackerman's *Frontier*. In this chapter, I consider how his life and work up to that 1963 lecture directly contradict the plea for an objective, inevitable, and rational science. The call for objectivity relied on appeals to intuition and subjective creativity. Geographers' subjective creativity can flourish only if the process of "mental structuring" of geographical thinking receives the same kind of social status and material support provided to other sciences; such support is by no means inevitable. Moreover, the essence of analytical, process-oriented conceptualization at the heart of General Systems Theory was premised on a fatal conceit of a singular, unitary rationalism—obscuring the exploding arrays of multiple, conflicting, and interacting rationalities evolving in the cybernetic age.

The Passions and Poetry of Science

The first contradiction involves the matter of objectivity. Ackerman's enthusiastic embrace of the "universality of scientific method" implied clear, unambiguous communication through the objective language of technocracy and mathematics—and yet his address is studded with digressions of polyvalent, nuanced ambiguity. Particularly in the footnotes, Ackerman unleashes the literary aspirations of his freshman arrival at Harvard. Not only does he build his argument on Weaver's portrayal of scientists as poets, musicians, painters, essayists, and philosophers; he also cites the zoologist Marston Bates's declaration that "science is only one of man's approaches to the understanding of the universe and himself. By understanding, I mean trying to make sense out of the apparent chaos of the outer world in terms of the symbol systems of the human mind. This might be considered the function of all art; and in that case I am led, half seriously, to call science the characteristic art form of Western civilization."[7]

But Bates is quoted in order to dispense with a criticism

from the cultural geographer William L. Thomas, who had sent a letter reminding Ackerman that geography was also part of the humanities. "This paper makes no pretense to coverage of all the ways in which geography may be viewed," Ackerman responded; "It discusses geography as a science."[8] Nevertheless, the overall argument of *Frontier* is for a General Systems Theory of science that makes *every* pretense of universal coverage. Repeated calls for objectivity, moreover, directly contradicted some of the most important achievements of Ackerman's own career. As Trevor Barnes and Jeremy Crampton document, Ackerman's initial work at COI/OSS conformed closely to the directives of Hartshorne, the "enforcer of objectivity," who safeguarded "scientific standards of language, truth, and logic"[9]—maintaining what Hartshorne described as a "clinical attitude" avoiding "crime[s] against objectivity" involving "hortatory and value words and phrases." "Expressive" description was to be avoided in intelligence reports in favor of "terseness and clarity."[10] But clinical writing stripped of expressive elements too often became the dry, encyclopedic, mind-numbingly meaningless compendium that was giving human geography such a bad name both in military applications and the lower stakes battles of academic turf wars. Effective communication of an engaged, relevant human geography required an approach that was simultaneously more systematic and more specialized—and more explicitly geared to the strategic priorities of the military enterprise. The truth of geographical science could not tell its own story. It required the unique blend of analytical flair and rhetorical skill of the new generation of brilliant young human geographers like Ackerman and Ullman, who were actively "redefining the character of geographical intelligence, moving it from the butterfly collection of disconnected facts, to something more targeted and incisive, conceptually chiseled by a sharper, more directed thematic purpose."[11]

Ackerman's *Frontier* call for objectivity echoes with bizarre ironies. While he appeals to the authority of Warren Weaver to stake a claim for the distinctive "mental structuring" of geographers, Ackerman draws on the prestige and reputation of Weaver the scientist—yet another mathematician who helped revolutionize antiaircraft fire control systems, who led a team for the National Academy of Sciences analyzing the genetic consequences of the atomic age, who helped launch the Rockefeller Foundation's support for the modernist "Green Revolution" in agriculture, and who helped create the Shannon-Weaver paradigm for information and communication theory. Yet Ackerman cites a speech delivered at a momentous transition in Weaver's life, shortly after his retirement from thirty years at Rockefeller. Weaver, who had written a short piece in 1949 "providing the original stimulus to the field of machine translation"[12] that would inspire generations of advances in artificial intelligence, machine learning, and natural language processing, was suddenly able to devote more time to family, reflection, and a mixture of other "major and minor enthusiasms." He gave talks on the interrelations between science and religion (for him, there was no conflict). He began work on a memoir, in which he judged that in his youth he "had a good capacity for assimilating information" but that he "lacked that strange and wonderful creative spark that makes a good researcher," and thus realized that "there was a definite ceiling" on his possibilities as a mathematics professor.[13] And he indulged longtime "minor enthusiasms" of collecting—especially *Alice in Wonderland* and "Carrolliana." Weaver collected

160 different translations in forty-two languages and wrote an entire book—*Alice in Many Tongues*—grappling with the challenges of translating a text defined by puns, parody, verse, and joyous twists of meaning. After Weaver's death his Carrolliana collection—one of the world's largest—went to UT–Austin, along with forty years of correspondence Weaver had accumulated in his efforts to track down everything written by the Reverend Charles Dodgson. And half a century after Ackerman relied on Weaver's scientific credentials to define a unique positivist "mental structuring" of geographers, *Alice in Wonderland* is now routinely used as a text corpus for the refinement of neural networks, machine learning algorithms, and even "virtual human" information retrieval agents with highly adaptive "linguistic, emotional, and social skills."[14]

GEOGRAPHY AND THE SOCIAL FORCE OF SCIENCE

The second contradiction involves the teleology of an inevitable geographical science. Reflecting on the half century of progress in his lifetime that had produced "a more profound change in our knowledge . . . than was achieved in all man's previous existence," Ackerman equates the modernity of the present "truly epochal period" with "the final world acceptance of science as a tremendous social force."[15] Ackerman's declaration may have been in tune with his NAS/NRC Committee colleagues and at least some of his Denver audience, but from our vantage point you and I can see the stark contrast with the contingencies of politics, personalities, and paradigms in human knowledge.[16] Social forces are social movements of strategy, tactics, and resistance; if science is a social force, it too can advance and retreat, and it can be resisted. Domains of knowledge with uncertain, contested claims to the privileged status of science are especially vulnerable to subversion and sabotage because science is made by human scientists in the context of the sociology and politics of the institutions of knowledge production. Students are allowed to receive formal credit and recognition for learning the distinctive mental structuring of a geographical imagination only if there is institutional authorization, and at Harvard Ackerman should have learned the bitter lessons of the failed appeals to tradition, the damaging legacy of Hartshorne's *Nature*—the presumption that "if geography is not given its due or is not in command of all of its rightful intellectual terrain, this is the result merely of illogic, which of course will be corrected once carefully explained to the academic powers that be."[17] Nothing was inevitable in the Harvard saga: logic and science were all mixed up with passion and prejudice. For one generation of geographers, the Harvard disaster was widely interpreted in terms of personal, idiosyncratic circumstances. Derwent Whittlesey was gay, and it was an open secret in Cambridge. He lived with Harold Kemp. They called their apartment, just at the edge of campus at 20-A Prescott Street, "The Loft." Whittlesey typed many of his personal letters on stationery with "The Loft" in an Old English font right at the top, center. "Whittlesey was the man, Kemp the woman," Preston James told Neil Smith in 1982; "They always had pink-faced undergraduates over there. They took them to the opera."[18] Smith, leading a new generation of critical reassessments of the history, notes that in the "gathering cold war hysteria," homosexuality "was deemed by many to be as un-American as communism, both unnatural and dangerous threats to a manly capitalism."[19] But generalized homophobia cannot be easily equated with allegations of institutionalized pedophilia, and there were other reasons why the circumstances had taken a severe toll on Whittlesey's credibility. Whittlesey—Whit, to those who knew him well—was desperately in love with Harold, and was his constant, loyal advocate: they had come to Harvard together after Whit had tried and failed to get a teaching position for Harold at Chicago. At Harvard, Whit never succeeded in getting Kemp anything more than a limited-term instructor appointment—but he never stopped trying, and he burned all of his institutional political capital in his attempts to promote a scholar who was widely considered less than mediocre. Kemp never proceeded beyond a master's degree, and even amid a faculty with little charisma, his introductory human geography course was regarded as a boring, anecdotal "white shoe" or "gentleman's C" course.[20] Kemp also seems to have had an especially abrasive personality.[21] More broadly, the field's most explicit attempt to unify its internal differences while joining a scientific mainstream—environmental determinism—had been thoroughly discredited as early as the 1920s, and in the ensuing backlash the severing of ties with the earth sciences and historical, process-oriented thinking led Carl Sauer to call the interwar period the "Great Retreat."[22] There was also the matter of Alexander Hamilton Rice, a onetime explorer of the Amazon whose wife had effectively purchased him a professorship with a million-dollar gift to establish an Institute of Geographical Exploration; Rice, an entrepreneurial self-promoter more than a scholar, commuted from Newport, Rhode Island, in a chauffeured Rolls-Royce and was an embarrassment to the other Harvard geographers.[23] Further complicating matters, the million-dollar endowment had come to Harvard after a rejected offer to the American Geographical Society, contingent on the ouster of AGS president Isaiah

Bowman to make way for Rice. There was thus quite a bit of personal history involved when Bowman, then president of Johns Hopkins, was appointed to the Ad Hoc Committee responsible for handling Ackerman's promotion. While the committee voted seven to four in favor, the chair of the Geological Sciences Division maneuvered behind the scenes to question the legitimacy of human geography, and Bowman was alternately incoherent, inconsistent, and reluctant in his attempts to persuade Harvard president Conant of the value of the field. It could have, and should have, worked out differently. Bowman, even more prominent than Hartshorne, was good friends with Conant: they had served together on a number of government committees; Bowman had influenced Conant's views on the role of (physical) geography in the establishment of the National Science Foundation; and a year later they were both part of a Top Secret Defense Department committee (along with Eisenhower and John Foster Dulles) charged with deciding how much the American public would be told about the military's atomic, biological, and chemical weapons research.[24] But for Bowman, human geography was too easy and could not match the scientific integrity of physical geography. Bowan regarded the human geographers at Harvard as nothing more than an "intellectual kindergarten."[25]

We can never know exactly how much of this wider context was clear to Ackerman at the time—and indeed, the sensationalized "rumors and legends" of Harvard geography's demise have always been defensive, obscured by "a fog of mythology" and personalized anecdotes full of "troubling discrepancies and contradictions"[26]—but we do have some evidence of how Ackerman tried to make sense of what was happening to him. "Whit," Edward wrote on a sheet of stationery from the University of Illinois in August 1948,

"This is my only copy of my resignation letter, in case you need to refer to it."[27] Ackerman began his letter to provost Paul Buck with standard clinical formality:

> I submit my resignation from the Faculty of Arts and Sciences of Harvard University, to be effective as of August 31, 1948. I am sure that it is not necessary for me to give reasons for my resignation.

But then Ackerman gave some very specific, damning reasons:

> As a loyal alumnus of Harvard I have sought to convince myself that the geography decision was for the best interest of the University. I must admit that it has been impossible for me to do so. Careful listening and much thought have only left me profoundly disappointed in the University's actions, both from an intellectual and an ethical point of view. That is not because of inconveniences which the decision may have caused me, but because the procedure and justification were disturbing in the implications they carried about American educational administration.

The point here is not for us to lament Ackerman's "inconveniences": his cover note to Whit was penned on Illinois stationery because just a few weeks previously he had turned down an offer to be head of the newly established Geography Department at Urbana, and he had been considering other good offers from Chicago, UCLA, Wisconsin, and Northwestern. Rather, the point is for us to appreciate that even being "so close to the events," Ackerman was able to see beyond the personal stories of "heroes and villains" as he tried to understand the wider political meanings of an institutional decision to kill off a discipline.[28] The editors at Harvard's student newspaper, the *Crimson*, had described the decision as an "Academic War over the Field of Geography," and Ackerman wholeheartedly agreed:

> It is disquieting to learn that the model American university is less democratic and less international in its outlook than present-day universities in some former autocratic enemy lands. It can hardly be considered a hopeful sign for realization of American ambitions to set a world pattern for democracy and international-mindedness. In that light the decision may be considered to detract from the University's usefulness in national life, a fact which we all can regret, and which even the undergraduates have recognized with dismay.[29]

In the ashes of a global war, there is no gentle, polite interpretation of such words.

Ackerman's rage over the Harvard decision helps us understand the obsession with fulfilling geography's scientific destiny that animated his 1963 *Frontier*, but no account of the history can validate a modernist, teleological unfolding of an inevitable logic. Personalities and the political economy of knowledge mobilized to build paradigms can realign. Paradigms gained can be paradigms lost.[30]

"Mac Yields in Birth Control Row"

Conflicting rationalities yield volatile, unpredictable irrationalities. This is the third contradiction in the 1963 *Frontier*, where an analysis of scientific progress and technological breakthroughs in communication leads to the insights on the "cybernation" of "manipulating some aspects of society" amid the research advances of "understanding the process of human thought itself." This is where the wartime experience of Ackerman and his peers was retooled for a postwar world of modernization, development, and consumption—achieved through a foundational commitment to the scientific rationality that had so clearly proven itself in America's military victory. Now the same rational systems thinking could be applied to the big peacetime questions, such as, "What can we say about how people distribute themselves and their culture on the earth, given free choice?"[31] By strengthening the connections between geography and the behavioral sciences, General Systems Theory will enable breakthroughs in our understanding of information flows and the diffusion of technological innovation; this is the point where Ackerman cites Hägerstrand's brilliant theorization of the "propagation of innovation waves" through time, space, and societal networks and structures. Systems theory will also illuminate the latent regularities of psychological processes generating international political conflicts. Ackerman's vision here is an extraordinary planetary fractal of individual human minds: geographers could capitalize on their long association with history to recode global historical knowledge through systems thinking ("history would acquire scientific meaning through the dimensions given it by behavioral science")[32] while also delving into the opposite scale of geographical thought: "study of the brain is considered one of the most useful approaches to the study of systems generally."[33] "We may well reflect to what degree social reality reflects the structure of the brain," Ackerman adds in a footnote.

The breathtaking ambition of a systems rationality connecting social reality with the structure of the brain, however, is a matter of faith. Ackerman learned this the hard way in Japan, whereupon beginning his work as a technical advisor on natural resources he was given the standard Office Memorandum No. 214.4 on publicity, national security, and classified information. Publicity for release in the United States should be designed "to impress upon the public the importance of the occupation mission," everyone in the Natural Resources Section was instructed, because "popular support and understanding are best obtained when the public is well-informed."[34] Yet information, popular support, and understanding conform to no central limit theorem: it is deceptive and dangerous to confuse the technical procedures of sampling with the moral, ethical, and political complexities of sampling. In the domain of human discourse, sampling means selecting, excerpting, interpreting, and pulling words out of context—in other words, *editing*. Ackerman the legendary editor found himself on the receiving end of a harsh edit.

The trouble began with just a few sentences. Ackerman had spent twenty months on fieldwork and research in Japan in 1947 and 1948, working with the assistance of more than thirty scientists. Nearly all of 1949 was devoted to the preparation of a mammoth two-volume report assessing the prospects for Japan's natural resources and postwar economic recovery. *Japanese Natural Resources* was finally released under the imprint and authority of the U.S. Supreme Command at a press conference on December 30, 1949. The initial print run was twenty-five hundred copies, and Lieutenant Colonel Schenck told reporters that work was under way on a Japanese translation—expected to be the most elaborate and expensive publication ever prepared by the occupation forces. It's hard to tell precisely what happened at that press conference— the archives are incomplete and fragmented—but there was apparently enough instant controversy for *Stars & Stripes* to report that "Schenck emphasized . . . that 'the author is responsible for all conclusions and interpretations of fact.'"[35]

The controversial conclusions and interpretations came near the end of the report, where Ackerman tried to synthesize the massive, encyclopedic information on agricultural productivity, raw natural resources, import/export trends, and the prospects for restoring the economy at least to the standards of living attained in the prewar years of the early 1930s. It didn't add up. Ackerman's productivity calculations suggested that, with continued U.S. assistance, Japan could sustain about eighty million people at 1930s living standards. But the combination of a returning diaspora and a vibrant baby boom drove the population from about seventy-two million in 1945 to eighty-three million in 1949, and natural increase was running at more than a million each year. Ackerman then reviewed the suggestions of "many Japanese scholars and leaders" who had begun to focus "attention on emigration as their principal hope for stabilizing population." But the rest of the region is crowded, too, Ackerman wrote, before asserting that even a pro rata resettlement would accommodate only a small fraction of the island's natural increase. And then,

> In a world so crowded, migration even on a vast scale can at best only postpone the necessity of facing the issue while at the same time creating additional population problems in newly settled lands. A solution to the population problem, which is part and parcel of the resources problem, therefore should be sought for and obtained within Japan. The population problem was largely created by reduced death rates; in all humanity, its solution can hardly be sought

elsewhere than in reduced birth rates. However, if control of the birth rate is not achieved and Japan is left to its own devices, death control does appear likely to enter finally.[36]

Church officials and activists had been closely monitoring MacArthur's postwar recovery and modernization plans. The previous summer, a Catholic delegation in Tokyo had secured a promise from MacArthur that the issue of population planning was definitively off-limits for the occupation. Mac had been forced to distance himself from the inflammatory words of a previous technical advisor—the demographer Warren Thompson—who had warned of resurgent militarism and communism, and of cutoffs of U.S. food assistance, if Japan failed to slow its population growth. Thompson's apocalyptic predictions, as well as his dismissal of the notion that the Catholic Church or U.S. public opinion would oppose birth control measures in Japan, attracted international press coverage, "prompting Catholics worldwide to protest" a perceived occupation endorsement of birth control.[37] MacArthur was considering a run for the U.S. presidency and was concerned about losing the Catholic vote; at the same time, Mac and his generals were extremely sensitive to the possibility of any occupation policies that could invite comparison with Nazi abuses.[38] MacArthur quickly disavowed Thompson: "In order to prevent any misunderstanding and to eradicate any misconception, the Supreme Commander wishes it understood that he is not engaged in any study or consideration of the problem of Japanese population control. Such matter does not fall within the prescribed scope of the Occupation and decisions thereon rest entirely with the Japanese themselves."[39]

The Japanese had indeed been working through various decisions—as the Church watched closely. The Eugenic

Protection Law (Yūseu Hogo Hō) of 1948 had decriminalized abortion, and amendments in 1949 made Japan the first country in the world to permit abortion on socioeconomic grounds.[40] Paradoxically, though, contraception policies were to remain rigidly conservative for decades. In August 1949, when the Communist Party was the only opposition to a preliminary resolution on birth control in the Japanese House of Deputies, the Vatican newspaper *L'Osservatore Romano* warned that "legalization of birth control in Japan plays into the hands of the Communists and will permit the Reds to outnumber the anti-Communists" within thirty years.[41]

In early 1950, the Supreme Allied Command had issued the Ackerman Report with these dangerous words about "control of the birth rate" and "death control." Officers of the Catholic Women's Club of Tokyo, a pair of Catholic Women's Clubs from nearby suburbs, and the Yokohama Rosary Society wrote to MacArthur, objecting to the "heavily alarmist arguments for nationwide population control," which had "publicly violated your official ruling" regarding occupation neutrality on such matters.[42] As reports of the press conference and the release of *Japanese Natural Resources* spread through the wire services (with the first of several articles appearing in the *New York Herald-Tribune* on December 31), so did the condemnations and controversy. The Reverend Patrick O'Connor presented the Catholic Women's Clubs' letter to American audiences in a story in the *Boston Pilot*. Father William A. Kaschmitter of Cottonwood, Idaho, the American editor of the *Catholic News Agency* and a long-serving missionary in the Far East, declared it "reprehensible and disgraceful for men of one nation to try to mastermind another into artificial birth control which was well described by Theodore Roosevelt as race suicide."[43] The *Denver Register* ran a cartoon showing a big American

General, labeled U.S. OCCUPATION AUTHORITIES, leaning down to a small, hungry man labeled JAPAN, who holds an empty rice bowl (ECONOMIC PROBLEMS); the general offers a present on a pillow—a large sword, BIRTH CONTROL. The cartoon's title: "Dishonorable Hara-Kiri."

MacArthur retreated. He halted distribution of *Japanese Natural Resources*—only fifty of the twenty-five-hundred print run had gone out—and removed the imprimatur of the Occupation Command; the book would be turned over to a private publisher.[44] Before transferring the book, though, MacArthur ordered the "offending passages deleted." A correspondent in Tokyo sent a wire to Chicago and New York on February 8, with a dramatic lead: "Catholic women in Japan were 'completely satisfied' Wednesday at MacArthur's unconditional surrender of birth control as solution to Japan's population problem."[45] Lieutenant Colonel Schenck was "severely reprimanded for allowing MacArthur to become involved in a dispute with the Catholic Church."[46]

Stateside, Ackerman was left to defend himself from a distance while censoring his own work.[47] "I am puzzled by the attention and publicity given an obscure reference," Ackerman wrote in a press release; "references to population in any manner occupy less than one page of the 560 page report."[48] Ackerman further noted that he had never actually mentioned "artificial birth control." Ackerman claimed that "previous consultations with lay Catholics" led him to believe that his interpretations would not be objectionable "from the point of view of Catholic doctrine." "It was my desire not to offend any particular religious or cultural group," Ackerman wrote in his press release. Nevertheless, two other items in the "Douglas MacArthur Material" folder in Ackerman's archives shift the meanings of his words. One is a hand-corrected typescript of an original draft of the contentious page 528 of the book. One of the sentences was quite explicit: "The ultimate solution may have to come in the reappearance of the Malthusian death control" if other measures were not taken, Ackerman had originally written.[49] This sentence never made it into the first version of the book: someone in the editorial chain of command must have understood that the expression "Malthusian death control" would be provocative in a nation devastated by the world's first "final solution" nuclear annihilation, where contraception was being compared to "race suicide."

The second item is an unsigned, unlabeled single sheet of typed text. It is unclear who wrote and typed it: it could have been Ackerman, but more likely it was Schenck, or possibly one of the many men working under Schenck's command (some of whom had worked with Ackerman on the project). But surely Ackerman read this page, given its position in the file alongside handwritten drafts, press clippings, and several photographs of Ackerman himself on the ship en route to Japan. The page reads,

A counter-offensive depends on:

1. An authoritative statement by a high-church official (ruling by Rome?)

2. Official Church disavowal of Father O'Connor for non-factual reporting.

3. Publication in the US of your entire unexpurgated book.

4. A letter or so to Gen Mac would probably not help, but if letters are written to him they should be from top flight US citizens.

5. Mrs M., who has not seen your book, should tell you so in writing.

"Mrs M." was Mrs. E. P. Monaghan, president of the Catholic Women's Club of Tokyo.

The controversy did not end there. The text entered the headlines again after MacArthur denied a visa application for Margaret Sanger to visit Japan to deliver lectures on "planned parenthood." A military government source confided that "in view of pressure from Catholic Church groups it was believed impossible for General MacArthur to allow her to lecture to Japanese audiences without appearing to subscribe to her views."[50]

———

The details of this controversy are not what really matters here. As with the contradictions embedded in Ackerman's writing style, and the failure of Harvard's leadership to recognize the merits of scientific geography, there's a more subtle lesson for us. A stark contrast appears between what Ackerman was struggling to say and the smooth, universal rationalist cybernation of crystal-clear systems-theory communications that Ackerman would advocate in his presidential address to geographers in 1963. In late 1949 and early 1950, as Ackerman's words and defensive protestations bounced awkwardly around the world through press releases and wire services, the message couldn't get through. The words used to convey the pragmatic planning considerations of modernization theory for the American occupation were sifted and sampled with meticulous precision and refracted through the politics of Catholicism, General MacArthur's military authority, and anticipations of American public opinion. It's not what you say, it's what people hear: this is what the maliciously brilliant twenty-first-century Republican consultant Frank Luntz would have advised Ackerman, had it been possible

(Luntz was not yet born).[51] That moment in history—the late 1940s aftermath of global war, and the sudden simultaneity of technological advances of applied sciences in an American-led international order—would fundamentally alter the small, precarious discipline of geography. This period saw a sudden spread of metaphors from the physical sciences that shaped the emergence and public reactions to the vastly expanding possibilities of the information and communications revolution. Entropy and the second law of thermodynamics, in particular, conditioned the cybernetics of Norbert Wiener and the informational paradigm of Claude Shannon and Warren Weaver—transforming the conceptualization of knowledge from a "cognitive state of being" into a new scientism focused on information detached from content and semantics, "disembodied from human life" yet pervasive, ubiquitous in modern society.[52] Information and communications theories, nurtured and distorted by the military-industrial complex, conditioned the postwar development of all the social sciences. While historians of science now regard the societal effects as most pernicious in the fields of psychology and economics, a case can be made that even more far-reaching ontological dangers occurred at the nexus of history and geography. This involves the understanding of time and space, and so in the next chapter we consider how Ackerman's mental structuring produced a powerful yet dangerous new kind of worldview.

CHAPTER 4

Militant Neo-Kantianism

In September 1963, the executive officer of one of the leading scientific think tanks in the United States delivered the presidential address to the annual conference of the Association of American Geographers. He implored geographers to embrace the neutral, rationalist objectivity of the language of mathematical statistics; to deploy General Systems Theory to unify all the scattered, fragmented traditions of the field into an inevitable, emergent scientific geography focused on "the vast, interacting system comprising all humanity and its natural environment on the surface of the earth"; and to advance the frontiers of informational "cybernation" and behavioral-science breakthroughs in "understanding the process of human thought itself."[1] And yet Edward Ackerman's declarations contradicted some of the most important professional and personal lessons of his life. These contradictions—the repeated failures of neutral, objective discourse, the catastrophe of believing in the inevitable revelation of geographical science, the colliding irrationalities of an informational world—have a significance far beyond the curiosities of personal stories about Ackerman.

Reading Ackerman's archive reveals a glimpse of the mental structuring of an entire generation of geographers at the dawn of the quantitative revolution amid the consolidation of the military-industrial complex. That mental structuring can never be reduced to any simple functionalist account—there are innumerable historical and geographical contingencies that give rise to the individual and collective dimensions of consciousness formation in contemporary capitalist society[2]—but neither can we ignore the structural facets of "historically specific social imperatives" of knowledge production.[3] The "motivations of individual geographers" must always be understood in relation to "the structure of their discourse,"[4] as well as the wider constraints of the political economy of knowledge. In urban and regional planning in the late twentieth century, for example, the funding imperatives of Anglo-American Fordism coalesced with the scientific logics of systems theory to produce a durable ideology that "directly transcends all that is historically concrete and specific," creating a "transhistorical" but "empirically vacuous" methodology.[5] Yet the essence of systems theory is *control*. This drive for control is the real meaning of the "mental structuring" Ackerman positions as the juncture of historical inheritance and amnesiac scientific futurism.

Ackerman's *Frontier* thus represents a disciplinary manifestation of what Michael Curry has called the "architectonic impulse"—the visionary passion to ignore all contradiction and ambiguity in order to "create an ordered, hierarchical system" of comprehensive explanation, . prediction, and control—at its most pure.[6] Curry traces the concept (architectonic) and the word to an 1877 critique of Kant, arguing that it captures a powerful, recurrent underlying tendency in geographic thought from Vidal de la Blache's *genre de vie* (mode of life) to Chicago School human ecology and structural functionalism to the mental mapping innovations of post–quantitative revolution behavioral geography.[7] Such tendencies also shaped the development of computer cartography and the long-running struggle from the 1970s through the 1990s to refine the technologies of geographical information systems (GIS) and then to pursue a more ambitious disciplinary project of geographical information science (GIScience).[8] In this chapter, however, I suggest that we should push Curry's analysis further across time and space. The concept and the term "architectonic" are explicit in at least one translation of Kant himself. They are implicit and foundational in geographers' work in Washington, D.C., during the Second World War. And they are today implicated in the transformation and selective dehumanization of human thought as architectonic logics are encoded into the increasingly automated systems thinking of big data, smart cities, cloud computing, and the Internet of Everything.

KANTIAN "TOTALITIES OF GEOGRAPHY"

For geographers, one of Hartshorne's major contributions was to revive a long-overlooked intellectual justification for geography's "position in relation to the other sciences" in the work of Kant, "one of the great masters of logical thought."[9] This appears in the *Gesammelte Schriften* (Collected Works) in a set of notes and outlines for an enormously popular course on physical geography that Kant offered at the University of Königsberg for forty-eight semesters between 1756 and 1796. The precise lineage of authorship has been contested for centuries. Kant had to obtain special permission to teach a subject for which there was no recognized, existing textbook. His lectures were based on voluminous personal notes he developed over the years. In *Anthropology from a Pragmatic Point of View*—a course manual published in a first edition in 1798, a second in 1800—Kant remarked in a footnote, "As for physical geography, it is scarcely possible at my age to produce a manuscript from my text, which is hardly legible to anyone but myself."[10] In contrast to the anthropology course text—which Robert Louden notes is unique among all of Kant's books because "it is the only one for which a virtually complete hand-written manuscript (prepared by Kant) still exists"[11]—Kant never published his geography course manual. He was infuriated when the publisher Gottfried Vollmer in 1801 began publishing a set of volumes assembled from the thriving black-market trade of notes taken by Kant's students. The first volume purporting to be Kant's geography course was based on student notes from 1778, 1782, and 1793. Kant scrambled to work with a former student, the publisher F. T. Rink, to release his own "authorized version." But by that time he was too frail to supervise the process; Rink was therefore responsible for many decisions on what to include, in what order, and in what writing style. Hartshorne reviews the authenticity disputes in a footnote that exceeds a full page of text, but ultimately concludes that we can trust the inherited materials on the "fundamental interest" of geography's relation to other fields.[12]

A crucial passage appears early in the published version of Kant's lectures on geography, shortly after he discusses the dichotomy in how elements of empirical knowledge may be classified—according to either "conceptions" or the "time and space in which they are actually found." Then, Kant reasons, "Description according to time is history, that according to space is geography. . . . History differs from geography only in consideration of time and area (*Raum*). The former is a report of phenomena that follow one another (*nacheneinder*) and has reference to time. The latter is a report of phenomena beside each other (*nebeneinander*) in space. History is a narrative, geography a description. Geography and history fill up the entire circumference of our perceptions: geography that of space, history that of time."[13]

For Hartshorne, this "filling up the entire circumference" view runs from Kant through von Humboldt and Hettner. It is Hettner's version—reality as "a three-dimensional space" defined by logical classification, time, and space[14]—that is the linchpin of Ackerman's wartime neo-Kantianism. This is because Hettner's Kantian "circumference" had been misunderstood as a justification for separating the two kinds of human geography that Ackerman had studied for his doctoral exams—regional *versus* systematic. Ackerman asserts that Hettner used the "unfortunate simile" of a series of plane surfaces parallel to the earth: systematic geography was concerned with the phenomena on a single plane surface, while regional geography studied "a limited section stricken through all of the parallel surfaces."[15] Ackerman, who had gained admission to a discipline dominated by regionalists only to find that the U.S. military was demanding systematic specialists, sought to transcend the dichotomy with a unified approach that he called "monistic geography." Parallel planes are only two-dimensional. What we really need is a "totality of geography" represented by a solid: "Regional geography is a vertical section, systematic geography a horizontal section of the solid,"[16] but the two are always interrelated, with any exploration tracing out a shape with multiple, "opposite faces" in a part of the totality.

At this point, it is worth acknowledging the present-day echoes of these concepts. Kant's "entire circumference" and Ackerman's military-intelligence global "solid" spring from exactly the same cognition that leads Mark Zuckerberg to form a technology consortium (Internet.org) that fuses the NASA-funded Gridded Population of the World database of "the geographical distribution of the human species" with information mined from Facebook's billion-plus user database to calculate that 85 percent of the world's people live within range of a cell tower with at least a 2G network; since "every human must be online," for the other 15 percent there will be a fleet of 747-sized solar-powered drones flying geosynchronously at sixty thousand feet.[17] Ackerman's monistic global Kantian solid also reappears in the portrayal of the low-light and thermal sensors of the U.S. 4th Special Operations Squadron's AC-130U gunship as providing a "'God's eye' view of the battlefield in almost all weather conditions,"[18] and in the "Collect it All" mantra of the U.S. National Security Agency—where in a single one-month period in the spring of 2013 the Boundless Informant program collected data on more than 97 billion emails and 124 billion voice calls from around the planet.[19] This is where the epistemological contradictions of Hartshorne's incomplete neo-Kantian idealism—conceiving knowledge "more as geographical terrain than historical process," and defining the discipline in terms of a central organizing concept (the region) that exists "only in our thoughts" untethered to any coherent materialist foundation[20]— is equally relevant to Ackerman in 1945 or 1963 as well as today's militarized surveillance capitalism. To the end of his life Kant also taught the "science of fortification" and provided astute "strategic forecasts concerning the wars in Europe,"[21] but only from a critical perspective in service to a broader theory of humanity's destiny of cosmopolitan peace.[22] Yet Hartshorne and Ackerman wind up theorizing a totalizing and militarized vision—and, as Derek Gregory demonstrates, the aspirations of militarized vision always remain hideously impossible: the "intimate entanglement of the technical and the human" creates inescapable, combinatoric contingencies of human and machinic errors masquerading behind the appearance of technical precision.[23] Cascading techno-social errors in the cognitive commodity chains of observation on the night of October 3, 2015, led the crew of that God's-eye AC-130U—a holdover from the Vietnam War upgraded with sensors and side-firing 25-mm and 40-mm canons and a 105-mm howitzer—to unleash an hour-long barrage of cannon fire circling over a Médecins Sans Frontières hospital in Kunduz, Afghanistan. At least forty-two people were killed, many burned alive in operating rooms. The NSA can vacuum the planet for billions of fragments of transmitted metadata, but the agency's notorious information-bulimia stupor is a perpetual reminder that understanding human communication—even or perhaps especially at the aggregate global scale—has little to do with positivist knowledge of a distant, external world in the style of Ackerman's OSS JANIS reports. It's about an entirely different scale, about intuition and the learning process of individual human minds.

Ackerman's revision of Hettner encoded the planetary scope of Kantian architectonic thought into modern human geography, but overlooked the *processes* involved in creating that worldview. Ackerman had been led astray, and so had an entire generation of spatial scientists. Half a century after the first publication of *Nature*, Neil Smith finally got Hartshorne to admit that all the Kant references were just superficial appeals to authority to secure geography as a coherent disciplinary identity. Hartshorne only began serious engagements with Kantian *philosophy* around 1970. And it is in Kant's philosophy—and the epistemological implications of Kant's understanding of space—where we find the true significance of a planet of billions of spatiotemporally connected thinking beings, where IBM estimates that two-thirds of the globe's humans keep a smartphone within arm's reach at all times.[24] Kant's comments on the position of history and geography in the classification of knowledge were only a means to an end, as Kant sought in his teaching to convey a "world-knowledge," a cosmopolitan philosophy integrating physical geography as knowledge about the world as an "object of external sense" and anthropology as an "object of inner sense."[25] In turn, this was part of a broader ambition to develop a theory of the possibilities and limits of human knowledge itself.

Kant sought to delimit the capacities for human reason that could escape the "antiquated and rotten constitution of *dogmatism*"[26] in theological metaphysics—in which Descartes and Leibniz used the notion of "pure intellect" to make claims about God and the distinct essences of mind versus matter. Kant argued that the theological philosophers "could not possibly know what they claimed to know about such things, because direct knowledge of a mind-independent reality exceeds the capacity of the human intellect."[27] At the same time, Kant sought to avoid the abyss of Hume's empiricism, which, if "unchecked" by a recognition of the boundaries of reason, could also succumb to a twisted obsession with sense-perception dynamics to "extend beyond its own domain in the world of nature" and underwrite "unjustified assertions about such topics as the free will of human beings and the existence of God."[28]

It is worth considering in some detail Kant's route between the extremes of Cartesian theological introspection and Humean empiricism. Kant accepts an empirical world of "things-in-themselves," but problematizes the relations between empirical and "pure" knowledge: indeed, "it is quite possible that our empirical knowledge is a compound of that which we receive through impressions, and that which the faculty of cognition supplies from itself."[29] Experience and perception create "impressions" in us, but these sense perceptions become coherent only through the order imposed by human sensibility and intuition. A gap opens up between an inaccessible *noumena*—pure entities that transcend all possible experience and sense perception (from the Greek *noomenon*, "the thing perceived")—and the *phenomenon*: a composite entity coproduced by an observing, thinking human Subject and a transcendentally real yet inaccessible perceived Object. The phenomenon is a human construct that "intercedes between the thing-in-itself and conceptual discourse, but is a product of them both."[30] Both, "although the one is not contained in the other, still belong to one another (only contingently, however) as parts of a whole, namely, of experience, which is itself a synthesis of intuitions."[31] "The Kantian dialectic," Smith explains, "thereby attempts to rescue the unity of Subject and Object from the empiricist and positivist resignation to duality; Subject and Object commingle, are interspliced, in the phenomenon, yet are at the same time distinct."[32]

The transmutation of "things-in-themselves" into phenomena is achieved through the fundamental human capacities of intuition. The human ability to organize and interpret experience and sensation is constituted by two forms of pure, a priori knowledge: space and time. Rejecting the Newtonians' conception of space as an external, empty receptacle for all objects and events, Kant argues that space is a "subjective and ideal" form of intuition, arising "from the nature of mind like an outline for the mutual co-ordination of all external sensations whatsoever."[33] For Kant, the "pure intuition of space" of human knowledges like geometry offered clear examples of the many kinds of "thought space" of conscious realities beyond empiricism—of cognitions that are "not things in themselves but representations of our sensuous intuition."[34] Human thought—and the contemplation of relationships of time and space—precedes and "co-ordinates" all sense perceptions of external things. Moreover, this system is all-encompassing: "It is impossible," Paul Richards writes, "for anything to appear outside the geographical scheme of things. All occurrences, past, present, and future, must have geographical location."[35]

External Architectonic

The all-encompassing coordinative function of human cognition embedded in Kant's conception of the "architectonic" requires that we consider two distinct perspectives: an external-world-focused knowledge versus the internal, self-awareness domain of human reason. While it is obviously a false distinction to treat these entirely separately, doing so allows us to appreciate the way Hartshorne-Ackerman distortions have been encoded into contemporary cybernetic practices.

For Kant, human thought precedes human experience in precisely the same way that an architect envisions what will eventually become a real, externally observable structure:

Whoever wants to build a house, for example, first of all conceives of the whole from which all the parts will afterwards be derived. Therefore our present preparation is an idea of knowledge of the world. We are creating indeed just such an architectonic concept [*Begriff*], which is a concept in which the manifold will be derived from the whole. Here the whole is the world, the stage on which we shall present all experience. Travels, and intercourse with people, broaden the extent of our knowledge. Each contact teaches us to know mankind but demands much time if this goal is to be reached. If we are already prepared by instruction we already have a whole, a framework of knowledge which teaches us to know mankind. Now we are in a position to classify each experience and to give it its place in this framework.[36]

Kant explicitly tells us to think architectonically: we must know the world in its totality in ideas before we can learn empirical details about the world, to "broaden the extent of our knowledge."

Knowledge of the world—of the physical geography description of the earth, and anthropology as knowledge of human beings—requires preparatory instruction that will "anticipate our future experience in the world, giving us, as it were, a preformed conception of everything."[37] This conception must organize "the objects of our experience *as a whole*." Human knowledge is "not an aggregation but a *system*; for in a system the *whole* is prior to the parts, while in an aggregation the *parts* have priority."[38] Kant invokes the function of the encyclopedia, to "produce an understanding of connections," in which "the whole becomes apparent only when seen in context."[39]

How do we see that context, to broaden the extent of our knowledge, to know the world in its totality in ideas? Kant admits that knowledge is gained by traveling, to have direct experience of place, to meet people. But he never really enjoyed travel, and he never fully trusted travelogue knowledge. His entire life of eighty years "bears witness to a necessity for thinking abnormally over the necessity of seeing."[40] He read everything he could get his hands on, and he often knew the details of distant cities and countries better than visitors who had grown up in those places—but to travel there would be a worthless, confusing distraction. From an early age he "forcibly closed his eyes and ears—the whole machinery of his senses. In spite of all inducements he never went further from Königsberg than a neighboring property, and even that he could not put up with for long, because all change in his surroundings disturbed his thoughts."[41] There is simply too much in the world to approach it solely through the limited and often unreliable channels of empirical sense perception: "This is not sufficient to enable us to know everything; as far as time is concerned, human beings live for only a short interval and can therefore experience only a little for themselves, but as for space, even if a person travels, he is still not in a position to observe or perceive a great many things. Therefore, we must necessarily have recourse to the experiences of others. But these experiences will have to be reliable, and for this reason written information is preferable to that passed on merely by word of mouth."[42]

Now it is important to recall that Ackerman, in his 1945 "Wartime Research" article, repeatedly cites Hartshorne but never mentions Kant. Yet a distorted form of Kant's architectonic thought works its way into Ackerman's analysis—and into geography's relations with the military-industrial complex and the quantitative revolution. To begin with, Ackerman accepts that seeing the world through secondary sources takes priority over seeing a small part through one's own eyes. Although he had spent a great deal of time in fieldwork as a student, military necessity meant that at Harvard he was teaching hundreds of officers the geography of Japan long before he actually went there. This was the kind of knowledge at a distance that would flourish in the quantitative revolution, as field observation was replaced with mathematical modeling, statistical manipulations of census data, and, eventually, after "the little beeps of Sputnik announced the possibility of a new way of seeing and sensing our planet,"[43] satellite imagery. In the latter case, the observational scale of technologically enhanced empiricism would be given a new scientific name—remote sensing—by Evelyn Lord Pruitt (1918–2000), the geographer at the Office of Naval Research who helped fund Ackerman's NAS/NRC Committee on Geography. ONR also funded a large cohort of the doctoral dissertations that decisively shaped and diffused quantitative revolution innovations throughout the discipline.[44]

MILITARIZING THE TECHNICAL MIND

But Ackerman goes further, in a subtle and dangerous move. Perhaps reflecting the considerable authority he had exercised over the research and writing of so many other men at OSS, and perhaps in his desire to stake out an independence from Hartshorne, Ackerman advances his project for a "monist" geography of highly specialized systematic expertise by attacking the very social structure of American interwar regional geography: the individual human geographer, doing fieldwork over many years to learn as much as possible about a carefully defined region, and then writing the results as individually authored interpretations—often with a great deal of personalized eloquence and subjectivity. "Despite many illustrations of the geographic complexity of even the smallest area," Ackerman argues, "we persisted in believing that one person, even at a graduate student level, is competent to study, understand, and interpret all phases of a given region or locality. Nothing could be further from realism, or more deadening to scholarly progress in geography from this period henceforth."[45]

Ackerman repeatedly lampoons regional geography's "holistic bias," its tradition of "academic individualism," and its view of the world as a mosaic—"with a potential student assigned to describing each tile." Wartime service, where he had been part of a team of geographers in military intelligence exponentially larger than ever seen in any university setting, taught Ackerman that a unified, monistic geography could be achieved only through a cooperative, team approach to "systematic dissection and analysis." "Group research leads to understanding of the whole never reached by individual effort," Ackerman asserted, proposing that a systematic division of labor would support "much more efficiently concentrated effort" while also making more productive use of research workers with the "technical mind"—"the people whose understanding of details much surpasses their abilities in analysis, interpretation, or invention." Geography has attracted pretty much the same share of students of the technical mind as other fields, Ackerman argues, but the holistic and regional bias of our field had inappropriately forced them to try "analysis, synthesis, or interpretation" in order to prove themselves. The result was a burgeoning literature of superficial, amateurish regional studies. We should follow the path of fields like physics and biology, where a fine-grained division of labor efficiently allocates technicians to the "great deal of mechanical work of the mind and eye" that needs to be done, without requiring them to "act as interpreters of the totality of the field."[46] A systematic, cooperative structure for geography would create a setting "where a task can be found for every intellect."[47] It would facilitate more opportunities for "cross-fertilization," and enable the discipline to cope with the complexity of the real world: "If we are willing, as many of us now are, to admit that the intricacies of any important region are too much for any one student to handle perfectly, and are not to be satisfied with a scientific objective short of perfection—then systematic geographers working cooperatively on a regional problem are the only possible heirs to the individual regionalist."[48]

This all sounds so logical, so reasonable. These could be the words you hear today from almost any dean, provost, university president, or program officer administering applications for grants and contract support. And of course that's exactly what Ackerman was living, describing, and eventually implementing in his work in Washington, D.C. The ideas are now so commonplace, woven so deeply into the taken-for-granted worlds of academic bureaucracies, buzzword-bingo corporate boardrooms, and Silicon Valley promotional pop-culture mottos—with Amazon's Mechanical Turk, behavioral research "is moving from the lab to the cloud,"[49] and Google's happy crowdsourcing mantra reminds us that nobody's as smart as everybody!—that it's easy to overlook the horrific contradictions. The essence of the Kantian "author function" of modernity—the "epistemic structuring of the world by a human actor"[50] presenting a totalizing scholarly argument through the written word— is deployed to convince readers of the futility of individual understanding of even the tiniest part of the world. Yet the architectonic impulse of a God's-eye groupthink remains, yielding a FrankenKant epistemology of all-encompassing space-time unchastened by any reflexivity regarding the intricacies and limits of human knowledge. Interwar regional geography was and remains easy to satirize for its intense localism and its barn-taxonomy descriptive obsessions, but at least there is one redeeming outcome from dividing the world into a mosaic and assigning a potential student to describe each tile. The student almost always develops a sense of responsibility, a mental structuring reflecting the distinctive view of the world from a certain, localized perspective, and often a long-term reverence for the meaning of place "as an event in human consciousness"[51]—as a simultaneous "heightening of self-consciousness" that coevolves "with the progressive partitioning of space."[52] Cooperative systematic research optimizes the efficient allocation of mechanical work of the mind and eye to technician minds unconcerned with the totality of the field, but of course this is a militarized mode of knowledge. It presumes and reinforces obedience to command authority. It relies on, and reproduces, hierarchical thought. Thus it is no small coincidence that some of the most innovative quantitative revolution geography funded by ONR applied meticulous mechanical mind-eye work to the U.S. urban landscape to verify the scientific elegance of Walter Christaller's theories of nested hexagonal hierarchies of human settlement—while avoiding the bibliobiographical details of Christaller's work on the Nazis' *Generalplan Ost* for the "resettlement" of conquered territories.[53] For the new spatial scientists, what mattered was a theory "abstracted from any particular place and any irrational purposes to which central place theory might have been applied. . . . It was the theory itself, not Germany, not Christaller, that contributed to a rigorous urban geography."[54]

It is crucial that we reflect carefully yet creatively through these connections. We must think across the false

Hägerstrandian time-space separations between now and 1945 and 1963 to consider what Ackerman offers as a model of geographical thought. The world can be known to perfection. It can be known in its entirety, with a globe encompassed by a series of Hettnerian parallel planes that fuse together into an all-encompassing planetary atmospheric solid that locates all possible observations in Kantian time and space. Instead of dividing the world into a Hartshornian mosaic of local knowledges, we should slice through it systematically, so that we can become specialized in precisely defined phenomena. We should approach everything, including ant colonies and pretty girls, as complex systems that can be broken down into constituent, interacting elements. We can use mathematical statistics to analyze all of these interactions as systems manifesting universal principles governing human as well as physical processes. But even as a quantitative systems theory allows the entire world to be known with perfection, none of us can do it alone. None of us, working individually, can know even the tiniest region or locality in all its complexity. So we should not even try. We should work collaboratively, making the most efficient use possible of those with technicians' minds. And on the advancing frontiers of a universal science, cybernation will continue to remove individual decision-making while revealing the process of human thought itself.

INTERNAL ARCHITECTONIC

All of this sounds a lot like today's proclamation that digital and biotechnologies are the "new frontier," the next industrial revolution creating the "smart cities" of an urbanizing planet.[55] This is the promise of Google's self-driving cars, Amazon's one-click drone deliveries, the GoFundMe and Kickstarter crowdfunding campaigns for your favorite causes, and the artificial-intelligence-enhanced scientific revelations of big data. As the algorithmic aggregation of social observation and measurement comes to pervade more aspects of life, a socially networked age begins to transform the entire spatiotemporal context in which individual sense perceptions and transcendental cognitions are translated into externally observed behaviors. There is literally a world of difference between saying that neither you nor I can completely understand a part of the world and saying that we should not even try—that we should just reach for our smartphones to connect to the cloud, to the "complex sociotechnical systems that are embedded within a larger institutional landscape of researchers, institutions, and corporations, constituting essential tools in the production of knowledge, governance and capital."[56] Ackerman and Hartshorne overlooked Kant's warnings about aggregating fragmented individual knowledge, and they misinterpreted Kant's conception of the relations between external and internal worlds. This is understandable— the essence of neo-Kantianism is to outdo a Kantian corpus and intergenerational tradition that itself seems infinite—and it forces us to confront the question of whether consciousness determines our existence, or existence determines our consciousness.[57] Kant's *Physical Geography* is premised on a seemingly absolute ontology of idealism: in between his metaphors of the encyclopedia and house building, Kant declares, "Idea[s] are architectonic; they create the sciences. . . . What we are doing here is making an *architectonic concept* for ourselves, which is a concept whereby the manifold parts are derived from the whole."[58] And yet almost immediately Kant takes his thought in a very different direction, right before the history-and-geography "circumference" passage that is incessantly quoted to situate geography as a university subject. "We extend our knowledge through the testimony of others," Kant emphasizes, "as if we had lived through the world's entire past. And we increase our knowledge of the present through testimony concerning foreign and remote countries, as if we had lived there ourselves."[59] Kant is here offering a premonition of the embodied, situated, genealogical reproduction of human knowledge that Foucault would later theorize as the "author function," the "privileged moment of individualization in the history of ideas, knowledge, literature, philosophy, and the sciences."[60] Foucault's purpose was to challenge that privileged moment, to contextualize the historical, discursive, and intertextual constitution of centuries of modernity yielding varied forms of authorship. What matters for us at this point is the way Ackerman's "mental structuring" and his 1945 attack on regional, holistic geography obscured the very conditions of possibility of any kind of geography that could be called "human." Kant's way of introducing students to the ideas of physical geography and anthropology—knowledge of physical description of the earth and knowledge of human beings— is to connect experience and perception of an external world to (1) communications flowing through humans providing testimony that extends the self across time and space,[61] and (2) the possibilities and limits of human cognition itself. "We must note that every experience of another person is imparted to us either as a *narrative* or a *description*," Kant explains, defining the former as a history and the latter as a geography.[62]

This is the "strange empirico-transcendental doublet" that Foucault so admired in Kant's "Copernican" revolution in knowledge: humanity is "a being such that knowledge will be attained" in humanity "of what renders all knowledge possible."[63] Foucault identifies a divergence between two kinds of knowledge—those "that operate within the space of the body . . . studying perception, sensorial mechanisms, neuro-motor diagrams, and the articulation common to things and to the organism," focusing on the "anatomo-physiological conditions" of knowledge, versus the "historical, social, or economic conditions" of knowledge "formed within the relations that are woven" among humans.[64] In a valuable retrospective on two decades of the "Idealist Dispute in Anglo-American Geography," Michael Curry analyzes how this duality has coursed through the multiple generations of post-quantitative revolution challenges to Kantian and neo-Kantian conceptions of reality as constituted by the human mind.[65]

Several considerations are important here. First, Kantian spatiotemporal conceptions of knowledge rely on particular scales of geography to condition the process of learning about the world. Before broadening knowledge through travel, Kant explains, one first "must have acquired knowledge of human beings at home, through social intercourse with one's townsmen or countrymen."[66] Kant then includes a footnote describing the features of political and economic centrality of Königsberg that make it "an appropriate place for broadening one's knowledge of human beings as well as of the world, where this knowledge can be acquired without even traveling."[67] The counterpart for Ackerman, of course, was the axis of twentieth-century American cultural and political power between Boston and Washington during the consolidation of the military-industrial complex. Today, the mobile communications revolution has unhinged information flows from the urban and regional scale, although this delinking remains partial and uneven—with enduring synergies between local life and the informational space of flows.

A second consideration, however, involves the duality of internal and external architectonics and the matter of determination. In his history of statistical epistemology, Ian Hacking positions Kant as the last defender of "the doctrine of necessity" embedded in European theological scientific thought. Hacking begins his analysis of this doctrine with *Foundations of the Metaphysics of Morals*, where

> Kant took as a commonplace that it is "necessary that everything that happens should be inexorably determined by natural law." Free will became a pressing problem because of the conflict between necessity and human responsibility. One resolution broadly followed the thought of Descartes, who had supposed there are two essentially distinct substances, mind and body, or thinking substance as opposed to spatially extended substance. Everything that happens to spatial substance is inexorably determined by law. Hence all spatio-temporal phenomena are necessarily determined. That might leave room for human freedom, so long as it is mental. Kant's account of human anatomy was a sophisticated version of this. The two substances, spatial and mental, were replaced by two worlds, one knowable, one not. The free self dwells in an unknowable realm of noumena. Kant was so convinced a necessitarian that he had to devise an entire other universe in which free will could play its part.[68]

From our present vantage point, Hacking's critique at first seems a familiar warning of the dangers of an unhinged, idealism-fueled postmodernity, of an infinity of individualized worlds. Kant himself was actually quite careful in conceiving of this "entire other universe," warning that Cartesian speculations "back and forth" over "the traces of impressions remaining in the brain" is "a pure waste of time"; in this "play . . . of representation" the human subject/object is "a mere observer and must let nature run its course," because the self-observer "does not know the cranial nerves and fibers," nor do they understand how to put them to use for their purposes.[69] I would suggest that Hacking's diagnosis applies less to Kant than to his successors—especially a now-forgotten conservative defender of theocratic psychology who provided a crucial linkage between German neo-Kantianism, French positivism, and the transcendental idealism of nineteenth-century New England politics and education that shaped Isaiah Bowman's intellectual development.[70] Victor Cousin (1792–1867) was regarded as the "official head of French philosophy" from the Restoration through the 1850s,[71] and distorted and betrayed Kant's entire project in an attempt to defend Europe's dying medieval theological political order. Cousin developed a doctrine of "interior observation" premised on a nondeceiving God who allows introspective observation of the soul, yielding an "inwardly directed" counterpart to the external scientific observational revolutions transforming human understanding of nature.[72] "Reason is in man," Cousin proclaimed, "yet it comes from God. Hence it is individual and finite, while its root is infinite," such that "the truth is in us," an echo of "the utterance [of] the eternal word of God."[73]

The militant neo-Kantianism forged through Hartshorne and Ackerman's engagements with German geography and U.S. "American Century" modernity was clearly and consistently secular. Yet the fatal contradictions of interior-observation Judeo-Christian teleologies of human nature define the very essence of "American exceptionalism" underwriting the strange coalition of Protestant Evangelical antimodernity and aggressive, deregulatory technological worship that sustains conservative hegemony. Militant neo-Kantianism is a particular kind of conception of the monist "totality of geography" that Ackerman distilled from Hartshorne and Hettner. It is premised on the assumption that meanings, intentions, and understandings can be inferred from distant observation, with data extracted, abstracted, transmitted, reassembled, dissected, and analyzed in a strategically managed, optimized division of labor. The *militance* involves the control over the division of labor, the protocols of extraction and reassembly of information, and the power over decisions on how and what to infer; the *neo-Kantianism* is genuinely new, because Kant could be blamed for creating an infinite alternative universe only two centuries after his death, when his architectonic of space and time became a genuine possibility—a mundane, pragmatic, everyday possibility within the constraints and rhythms of daily life and decision-making—for a large proportion of the world's human beings. That proportion must be sufficiently large that it becomes realistic to take Bertrand Russell literally when he portrays space and time as "part of our apparatus of perception," in that half of the doublet Kantian phenomenon involved in giving order and coherence to the overwhelming chaos of raw, empirical sense perception. This half of the empirico-transcendental doublet "is not itself sensation . . . it is always the same, since *we carry it about with us*, and it is *a priori* in the sense that it is not dependent upon experience."[74]

Thus we now live among a proliferation of cybernetically mobilized *phenomena*—interspliced, coproduced entities of pseudo-empirical perceived realities and human cognitive engagements and experiences, an accelerating dialectical subject/object circuitry that, more and more, we literally carry about with us in our smartphones. These phenomena are not just externally observed, digitally recorded traces of decisions and movements; but neither are they pure analogue signals of internal human thought. They traduce the divides between analogue and digital, inner sense and outer sense, ideal and material, subject and object. The individualized, infinite alternative-universe idealism that Hacking diagnosed in Kant now coalesces with new scales of empirical sense perception and transcendental assertions. Late twentieth-century neuroscience finally made it possible, contra Kant, to know the cranial nerves and fibers and to understand how to put them to use, while twenty-first-century GPS-smartphone ubiquity has woven B. F. Skinner behavioral "operant conditioning" systems into thoroughly planetary networks. Silicon Valley's philosophy of "captology"—computers as persuasive technologies—blends behavioral psychology with interface design to "hack the human brain and capitalize on its instincts, quirks, and flaws" in an ever-accelerating decision environment of information overload.[75] Ray Kurzweil, chief of engineering at Google, portrays humanity in the very near future as a "hybrid of biological and nonbiological thinking," in which we will "basically put our brain on the cloud." "We are already integrated with our brain extenders," Kurzweil explains to the crowd in a recent public presentation, holding out his cell phone: "It's a gateway from my brain to the Cloud."[76] Kurzweil's current research frontier is edging us toward the point where "nanobots in the neocortex will communicate with the cloud," where "nanobots communicating with neurons" will integrate individual human cognitive capacities into globally networked collective intelligence. This is nothing short of a new neo-Kantianism, a new phenomenology in which the most dramatically unleashed infinities of idealism are poised between (1) the free will domains identified by generations of critics of neo-Kantianism, from Husserl to Hacking, as hijacked by late-capitalist postmodern neoliberalism, and (2) the relational materialism that runs from structural Marxism through contemporary streams of posthumanism and feminist new materialism.[77] The relations between external and internal architectonics are realigned, unhinged, and reconstructed through multiple scales of combinatoric, networked possibilities that are—from the perspective of finite individual human understanding—thoroughly and inescapably infinite. More than four hundred hours of video are uploaded to YouTube every minute. If ten important emails suddenly arrive in your inbox but you have time to glance at only three, the rules of combinatorics yield $10! / (10 - 3)! = 3,628,800 / 5040 = 720$ different permutations of demands on your time and attention; if it's fifteen emails and you go through ten, the permutation of sequential, path-dependent attentional possibilities exceeds ten billion. Some of the multiple scales through which combinatoric phenomena evolve are familiar—the "digital divide" dualism between real and mediated spaces of communications, from Gutenberg to telegraph, radio, cinema, television, and now Snapchat and WeChat—while others are more recent: once the line between real and virtual is crossed into the online world, there are ever-changing boundaries and scalar contingencies associated with various operating systems, apps, and algorithmic, correlational "filter bubbles" of customized, socially networked streams of information. The exponentiated network effects of the cybernation Ackerman glimpsed in a few key notes of the early postwar military-industrial complex are now ubiquitous, reconstituting the spatiotemporal essence of human perception and understanding. Militant neo-Kantianism is thus simultaneously scaled inward to the neurons (in a latter-day interior observation) and outward to the planetary scale. As Jim Thatcher emphasizes, "When mobile spatial data is taken to signal the presence of a phenomenon in time and space, there remains an epistemological leap from individual to data point: a leap from the individual and whatever motivations went into the creation of the data point—the humor, sarcasm, irony, and earnestness that accompany everyday life—and the purportedly quantitative facts the datum comes to signal."[78] As Thatcher demonstrates, "in a world suffused with data," subject formation and the constitution of identity create multiple contingencies of subject/object relations "influenced by and influencing assemblages of social, economic, and temporal factors that go into the production of space and identity."[79] In a world in which "mobile spatial data is increasingly the data through which individuals know and are made known to themselves and others," Thatcher explains, external observation and inference from the *doing subject*—capturing and analyzing movements and digitally rendered activities—must be complemented by inquiry into the "reflexive, self-eliciting subject" and the "reflexive interpretations made by individuals as they do and do not make use of mobile spatial devices in their daily lives."[80] There is some significance, then, in the way the anarchist philosopher of science Paul Feyerabend draws attention to a neglected passage in the *Critique of Pure Reason* in which Kant suggests, "Where common sense believes that rationalizing sophists have the intention of shaking the very fundament of the commonweal, then it would seem to be not only reasonable, but permissible, and even praiseworthy to aid the good cause with sham reasons rather than leaving the advantage to the . . . opponent."[81]

Kant is dubious of the enterprise of "maintaining a good cause" with "deceit, hypocrisy, and fraud,"[82] but he also understands that "disingenuousness," the "disposition to conceal our real sentiments" while keeping up appearances, has the effect of civilizing humanity by making public life "a school for self-improvement."[83] And, as Slavoj Žižek has observed in his "Hegel beyond Hegel" update of dialectical materialism, Kant does attempt to account for the possibility of freedom in a determinist universe with the "incorporation thesis": "I am determined by causes, but I retroactively determine which causes will determine me."[84] "Freedom" is thus "not simply a free act which, out of nowhere, starts a new causal link, but a retroactive act of endorsing which link or sequence of necessities will determine me."[85]

Retroactive combinatorial path dependencies thus percolate through computational Fourier analyses of the spatiotemporal subject/object reflexive, self-eliciting subjects of, at last count, 6.58 billion worldwide mobile users and, if we include the multiple phones, tablets, and other devices associated with individual humans, a total of 12.01 billion global mobile devices.[86] Strange mutations of Ackerman's monist geographies appear when General Stanley McChrystal, commander of Joint Special Operations Command (JSOC) between 2003 and 2008, delivers a TED talk—complete with a cute-cat image of the sort that is now a mandatory feature of the age of internet memes—reflecting on his experience of having "more connectivity than any commander in military history" in his Situational Awareness Room in Iraq. All the advanced technologies of connectivity seduce, deceive, and enslave us, McChrystal warns. We're drawn into the illusion that more information enables "an understanding, an appreciation, an empathy for what is happening to the people on the ground"—*true* situational awareness.[87] Meanwhile, a former JSOC drone operator explains how the High Value Targeting Task Force uses cell-site simulators to track SIM cards on the Geolocation Watchlist to direct drone strikes in Yemen, Somalia, Afghanistan, and elsewhere—often facilitating lethal strikes "without knowing whether the individual in possession of a tracked cell phone or SIM card is in fact the intended target."[88] One of the problems that precipitates the deaths of innocent civilians and bystanders, the source explains,

> is that targets are increasingly aware of the NSA's reliance on geolocating and have moved to thwart the tactic. Some have as many as sixteen different SIM cards associated with their identity within the high-value target system. Others, unaware that their mobile phone is being targeted, lend their phone, with the SIM card in it, to friends, children, spouses, and family members. Some top Taliban leaders, knowing of the NSA's targeting method, have purposely and randomly distributed SIM cards among their units in order to elude their trackers. "They would do things like go to meetings, take all their SIM cards out, put them in a bag, mix them up, and everybody gets a different SIM card when they leave," the former drone operator said. "That's how they confuse us."[89]

Not all cases of militant neo-Kantianism are so explicitly violent, but the logic is the same. Locating events as coordinates in space and time, decontextualizing spatiotemporal events from their embodied human experiences, intentions, and meanings—all of this tears apart what the historian R. G. Collingwood saw in Kant's co-constituted material realities and intergenerationally relived "inner lives" of human phenomena, all for the sake of efficient recombinant target-driven processing in a hierarchical division of labor.[90] McChrystal's situational awareness blends inner-reality Kantian idealism with Marxian materialism and algorithmic positivism when the "total field" of the spatiotemporally knowable is "cut up into an infinity of minute facts each separately to be considered" and then reassembled.[91] If we follow Collingwood's sympathetic critique of Kant and seek to get inside the inner lives of the phenomena connecting Ackerman and Hartshorne with McChrystal, it comes as no surprise to see all manner of hierarchical hijackings of human geography. When Andrej Holm was arrested in the summer of 2007 and rendered to solitary confinement on charges of "membership in a terrorist association," the evidence involved sophisticated technological inferences based on the research he and a colleague (Mattias Bernt) had undertaken on a direct action protest group. Authorities analyzed the scholars' writing to match keywords and punctuations with the declarations of protesters (use of the words "imperialism" and "gentrification," use of G8 rather than G-8); scholarly affiliation with a research institute was interpreted as "access to libraries" that could be used "inconspicuously in order to do the research necessary to the drafting of texts" of the protest group. Comprehensive surveillance yielded a flood of data enabling the prosecutorial dissection of Holm the urban researcher and his reassembly as "the brain behind the militant group," with mundane spatiotemporal points in his daily routine defined as "highly conspiratorial circumstances" when, for example, he went to meetings without bringing a cell phone.[92] It is a bittersweet legacy to see a parallax view of today's newest quantitative revolutions: a new generation of critical spatial analysts

is refining an astonishing array of methodologies for the
situated, contextual revelation of dynamic new worlds as
society coevolves with big data;[93] and yet geographers'
innovations are always surrounded, overshadowed, and
policed by an increasingly coercive infrastructure adapted
to the new scales of *planetary* observation of *individual
human phenomena*. We can see key episodes in the history of
geographical thought, then, not just in the Holm case—this
is the nightmare dystopia if Torsten Hägerstrand had worked
for the NSA—but in countless more mundane circumstances.
Mark Graham's extraordinary analysis of the globally uneven
geographies of billions of author contributions and edits
to Wikipedia,[94] for example, has a troubling counterpart in
the way advanced pattern-recognition software designed
to analyze brain waves was reconfigured to analyze "digital
fingerprints" in fragments of text compared with an evolving
cloud database of several hundred thousand publications,
three hundred million archived student papers, and more
than forty billion web pages across eighteen languages in more
than a hundred countries—such that a pair of students who
inadvertently reproduced passages of text from Wikipedia for
term papers in a semester-at-sea course could be convicted
of plagiarism aboard a ship touring the Mediterranean,
expelled from the program, and abandoned at the next port
of call. They wound up sleeping in the airport before paying
for their own way home.[95] One student's reaction to the
zero-tolerance pattern-recognition plagiarism policy—"it
is hard to remember what your thoughts are and what was
from Wikipedia"—has more than a trivial connection to our
consideration of Ackerman's thought frontiers and Kant's
assertion that ideas are architectonic. In a separate incident,
Emma Teitel, a student in her third year at Dalhousie, was

convicted of plagiarism in a creative writing course—not for any word-for-word match, but for a failure to cite the originator of a philosophical concept she used to analyze the history of cinema. Here is Teitel's offending passage: "Such a question is the stuff of a philosophical category mistake. For example, if a small child touring Halifax were to ask his mother and father, 'show me a university,' the parents might take him on a tour of Dalhousie, showing him all the different faculty and athletic buildings, and confused, he would still ask 'show me a university,' so the same concept applies to the question of film editing."[96] Teitel's description mirrors the example of "a foreigner visiting Oxford or Cambridge for the first time" in Gilbert Ryle's 1949 book, *The Concept of Mind*,[97] but Teitel makes a convincing case that she had absolutely no intent to plagiarize: "I had assumed that Ryle's phrase and its illustration had entered the academic vernacular, like Kant's 'categorical imperative.' I was wrong."[98] In the militant neo-Kantianism developed through Ackerman and Hartshorne's wartime experience, intention of the observed subject is irrelevant: what matters is the efficient management of vast datasets of spatiotemporal abstractions, an evacuation of intention and meaning from embodied, situated people and places, and reassembly into a scientific, corporate, or military hierarchy. What happened to Teitel and Holm is the fate that awaits us all, a culmination of a hijacked Kantian architectonic impulse backed by capital and code. It is not just the spatiotemporal paths of our lives in raw Cartesian coordinates that constitute the value pursued by Silicon Valley and DoD innovators; the real potential lies in the drive to draw inferences of the meanings and intentions of what Curry famously theorized as the "digital individual" acting and perceiving in spaces that are relational, transductive fusions of the physical and virtual,

concrete and cognitive, subject and object.[99] As more of the
world's information is rendered in digital form and curated
in corporate and surveillant databases enabling near-instan-
taneous analysis, humans near and far, living and dead,
are put into a macabre metaphysical encounter with Kant's
notion that we learn about the world through testimony from
others, as if we had lived through the world's entire past. Ryle
died decades before his words convicted Teitel, but Ryle's
motivation in writing *The Concept of Mind* was to challenge
the enduring "official doctrine" of Cartesian duality between
mind and body, mind and matter. Ryle introduced the concept
of a "category mistake" specifically to demonstrate the
absurdity of "the dogma of the Ghost in the machine," with its
polar opposition between a material existence in "a common
field, known as 'space,'" subject to readily observed Newtonian
cause-and-effect relations, versus a mental existence in
"insulated fields, known as 'minds,'" with "no direct causal
connections between what happens in one mind and what
happens in another."[100] This is the divide that seems to be
fast eroding in contemporary cybernetic implementations
of Ackerman's militant neo-Kantianism. In his eloquent
retrospective on the flourishing idealism turn in human
geography in the 1970s, Curry cites Sheldon and Susan Watts'
insightful critique of Leonard Guelke's influential "idealist
alternative" to the era's dominant positivist quantitative
geography. A major problem with Guelke's idealism, Watts
and Watts point out, is its reliance on Collingwood's attempt
to stretch a certain form of Kantianism across time—such that
the historian's primary task is immersion in texts, archives,
and discourses of an era in order to approach the possibility
of rethinking the thoughts of key historical actors. Watts and
Watts warn that "the geographer who attempts to rethink

people's thoughts" confronts "the argument put forward by the philosopher Gilbert Ryle that an individual's thoughts are, by definition, private and thus inaccessible to an outside observer."[101] Curry extends this critique, with a reminder that Collingwood's attack on Auguste Comte's positivist formula for correlating "the facts about human life"[102] relied on an arrogant presumption—that it is possible for a historian to become so deeply immersed in the texts, traditions, and circumstances of a historical period that one can understand the context and meanings that led to particular events. Curry further notes that when we look back across time, we have knowledge of the cause-and-effect consequences of actions that could never have been available to an era's actors: "we" in the present know things that "they" in the past could never have known.

Today, consideration of all of these dimensions of Ackerman's militant neo-Kantianism underscores two central points. First, data purporting to represent the very process of "mental structuring," of the nexus of outer (material) and inner (mental) space, are being comprehensively industrialized, from Facebook's 2.2 billion monthly active users to McChrystal's data-feed situational awareness room, Kurzweil's neuro-nanobot-cloud interface, and the artificial-intelligence advances used to surveil students' term paper submissions. From Cambridge Analytica's harvesting of personal data on 87 million Facebook users to calibrate "psychometric" targeted ads to the latest breach in which advertisers were given access to millions of private telephone numbers supplied by Facebook users solely for two-factor authentication, it becomes ever more difficult to maintain that individuals' thoughts in space and time are private and inaccessible to outside observers. Such a claim is less a philosophical universal than a political question of privacy and digital citizenship—even if, as Thatcher wisely emphasizes, the vast majority of big data inference succumbs to a fatally flawed epistemological leap from individual to data point.

This brings us to the second implication. Even the most serious epistemological flaws in data-driven observation become powerfully performative in a world suffused with seemingly infinite information. This is the true danger of what Žižek identified in Kant's incorporation thesis, in Hacking's attack on Kant's unknowable universe of *noumena*, because Collingwood's modernist optimism is reversed: it is possible for previous generations to have many kinds of knowledge that our own generation has distorted, destroyed, forgotten, or hidden in plain sight. Our current cybernetic information explosion has precipitated exponential, combinatoric expansion in the multidimensional spatiotemporal domains of possibility in which phenomena are created and communicated, modified and mobilized. Hence our technologically advanced, information-rich world of memes, conspiracies, and reincarnated Nazi ideologies.

———

Today, for so many scholars, entrepreneurs, and activists, there's an absolute, optimistic certainty in the answer to Ackerman's 1963 question, "Where Is a Research Frontier?" It's in the cloud. This is the new frontier of exponential expansion of data-driven possibilities that seem finally to have achieved von Bertalanffy's vision for General Systems Theory—"a new scientific doctrine of 'wholeness'"[103] in a wireless world of universalizing connectivity and code. Yet the frontier metaphor, so powerful in the geographical imaginations that shaped American identity in a world of colonization and dispossession, cannot be separated from danger and violence. The very essence of America, the historian Frederick Jackson Turner famously asserted, evolves through "a return to primitive conditions on a continually advancing frontier line," a dynamic wave whose outer edge is "the meeting point between savagery and civilization."[104] Frontiers are dangerous, and the Turner thesis, which came to be one of the most influential statements at the nexus of history and geography at the climax of social Darwinism in America, was quite explicit in the role of the "recurrence of the process of evolution" in the transformation of knowledge, society, and human nature itself. If we really want to take Ackerman's *Frontier* seriously, then, we must confront the savagery that drives its evolution. This will be our task in the next chapter.

The New Evolution of Geographic Thought?

With communication, evolution will take care of the rest. The ideas that are best adapted to the needs of society and the aspirations and career opportunities of the next generations of scholars will survive to shape the urban geography of the future.

—Brian J. L. Berry and James O. Wheeler,
Urban Geography in America, 1950–2000

We're asleep at the switch because it's not just a metaphor. In 1945 we actually did create a new universe. This is a universe of numbers with a life of their own, that we only see in terms of what those numbers can do for us. Can they record this interview? Can they order our books on Amazon? If you cross the mirror in the other direction, there really is a universe of self-reproducing digital code. When I last checked, it was growing by five trillion bits per second. And that's not just a metaphor for something else. It actually is. It's a physical reality.

—George Dyson, "A Universe of Self-Replicating Code"

The time scales for technological advance are but an instant compared to the time scales of the Darwinian selection that led to humanity's emergence. . . . In a long-term evolutionary perspective, humans and all they've thought will be just a transient and primitive precursor of the deeper cogitations of a machine-dominated culture extending into the far future and spreading far beyond our Earth. . . . It won't be the minds of humans, but those of machines, that will most fully understand the world.

—Martin Rees, "Organic Intelligence Has No Long-Term Future"

Geographic thought is evolving. With the passage of a single human generation, the eloquent premise of David Livingstone's *Geographical Tradition*—that geography's story, like all stories, is told by people, about people, and for people—has become an unlikely, tentative proposition that must be actively justified and defended. There's a remarkable, sudden convergence here of technological trends, popular stereotypes, and cutting-edge poststructuralist theoretical currents. With the speed and convenience of GPS, smartphones, and Google Maps, why do we need human geographers to tell us where things are? With the dramatic theoretical challenges of postcolonial theory, assemblage theory, and posthumanism studies, how can we cling to the arrogant Western Cartesian species conceit that there is anything individually distinctive or heroic about a human who seeks to think geographically? My purpose in this exploration has been to study a pivotal series of episodes in geography's quantitative revolution through the eyes and words of an important, overlooked human witness— in order to struggle through the antecedents of the current paradox of our field: in everything from the "software-sorted geographies" of daily consumer life to the precision drone-strike coordinates of cellphone SIM card tracking on the U.S. military's Geolocation Watchlist,[1] human geography is more popular and pervasive than ever before. Yet it is not always clear that we need human beings who work to become recognized as human geographers. In this chapter I highlight this paradox by analyzing how the scientific discourses of General Systems Theory that inspired Ackerman in the 1960s are guiding today's evolution of geographical thought and practice.

MONIST GEOGRAPHIES AND TWITTERBOTS

At the precise moment when the ready-retail convenience of ubiquitous connectivity has enabled an infinite chorus of stories about and for people at a scale that is planetary and personalized—Facebook has 2.2 billion monthly active users, who collectively upload more than 300 million photographs to the site each day and post 510,000 comments every sixty seconds[2]—it is becoming ever more difficult to discern the stories that are actually told by people. Facebook offers a series of apps that deploy facial recognition to automatically grant discounts to users who check in to camera-equipped bars and stores, part of an emergent wave of efforts to strengthen the data interface that connects "virtual" places with "real" physical-geographical coordinates.[3] At the same time, however, at least 83 million of those Facebook profiles are fake.[4] Bots account for 48.5 percent of all online traffic.[5] Three weeks before the 2016 U.S. presidential election, a professor of internet studies at Oxford estimated that fully one-third of the millions of tweets favoring Donald Trump were generated by automated Twitterbots (compared to only one-quarter of those favoring Hillary Clinton).[6] The very next day, Twitter and scores of other high-traffic nodes became virtually inaccessible thanks to the latest botnet DDoS (distributed denial of service) attack—orchestrated this time by Mirai malware infecting hundreds of thousands of internet-connected home surveillance cameras, baby monitors, and Wi-Fi routers with insecure password authentication.[7] A flood of investigative research by scholars and journalists further documents the deliberate, strategic deployment of aggressive pro-Trump "social bots" in the days and weeks leading to Trump's victory,[8] an astonishing upset that revealed, as the *New York Times* technology correspondent Farhad Manjoo put it, "Social Media's Globe-Shaking Power."[9] And for quite some time now, automated cybernation has also been reshaping formalized academic knowledge production, as a "Gutenberg Moment" transforms the social processes of scholarly peer review at the same time that pattern-recognition algorithms and machine learning innovations dynamically reconfigure the ways scientists search through citations and literatures to find the cutting edge of a research frontier.[10] Thus we should not have been surprised by a study undertaken by *BuzzFeed*, one of the wildly successful new hybrids of information, entertainment, news, and social media founded by an entrepreneur who studied with the feminist cyborg theorist Donna Haraway—and who published a scholarly article on the internet's role as a capitalist phenomenon allowing for "more flexible, rapid, and profitable mechanisms of identity formation" and dissolution.[11] The founding editor for *BuzzFeed Canada* studied the most widely shared news stories on Facebook related to the 2016 U.S. election and found that demonstrably false and misleading news stories "generated more engagement" than mainstream news in the last three months of the campaign. The top-performing fake election stories generated 8.71 million shares, reactions, and comments on Facebook, compared with only 7.37 million for the top-performing mainstream news stories.[12] The top performer, a fake news item announcing that Pope Francis had endorsed Trump's presidential bid, generated 960,000 of these engagement metrics—implying a remarkably efficient diffusion of a falsehood through an online population of several million. Facebook CEO Mark Zuckerberg, speaking at a technology conference shortly after the election, tried to dismiss the idea that fake news had any effect on the outcome as "pretty crazy,"[13] but his attitude was quickly challenged by journalists, editors, and other analysts who had learned painful lessons from trusting the scientific assurances of the "polling industrial complex."[14] After a long, grueling campaign dominating a year in which the Oxford Dictionaries named "post-truth" the word of the year,[15] analysts suddenly realized that the planet's most powerful cybernetic-military infrastructure—from DARPA to "smart bombs" and the NSA's "geolocation cells" pursuing metadata on cell-phone SIM cards to direct targeted drone-strike assassinations under the playful motto "We Track 'Em, You Whack 'Em"[16]—was now under the command of a "reality" television celebrity who first gained attention in the political arena by embracing the "Birther" lie believed by tens of millions in the networked outrage of a resurgent white-nationalist alt-right. Now, under an executive who "likes to surprise" and who "enjoys the worldwide speculation he sets off with his Twitter posts"[17] and who proudly declared an intention to deport millions while one of his surrogates justified a planned national registry of Muslims by citing the precedent of the 1942 internment of Japanese Americans,[18] state authorities are positioned to monitor ever more of the stories by, about, and for people. Humanity is becoming more thoroughly cybernetic, militarized, and surveillant, creating a bizarre and dangerous reincarnation of the planetary monist geography that Ackerman dreamed up while editing geographical reports at OSS.[19] The NSA has assembled data on at least twenty trillion interactions among U.S. citizens, and processes at least twenty billion communications events (emails, online chats, online browsing activities, telephonic metadata) each day.[20] The human storytelling process itself is now becoming its own militarized cybernetic research frontier: the application of artificial intelligence to discern authorship from linguistic evidence—"stylometry"—in criminal and terrorism investigations has achieved dramatic advances in the last decade; but there are concerns that human writers wishing to avoid identification can alter their writing to subvert algorithmic detection—giving rise to yet another new research frontier known as "adversarial stylometry."[21]

Seeking and Staying on a Research Frontier

To suggest that geographic thought is evolving, therefore, is not just a rhetorical flourish. It's a materially real and subjectively experienced velocity differential between (1) accelerated space-time compression wrought by communicative and linguistic transformation and (2) situated, embodied intergenerational changes in humanity. A widening divergence separates the relative speed of the reproduction of *human knowledge* versus the reproduction of human *lives, institutions*, and *social relations*. Ackerman caught a glimpse of this gap at the very end of his *Frontier* speech, when he reflected on the "most exacting task" of "seeking and staying on a research frontier." "I believe the time is near," Ackerman told his audience in Denver, "when postgraduate training and a second doctoral degree may be the price for reaching a research frontier."[22] As Ackerman spoke these words, his colleague Evelyn Pruitt was administering the ONR Foreign Field Research Program that was becoming so important for a new generation of geographers; of the 432 applications received between 1955 and 1966, 97 were funded (84 for doctoral students).[23]

Roughly 20 percent of each year's undergraduate applicants to Harvard in these years were gaining admission. More recently, Harvard's admissions rate has been marching down, from 9.7 percent for the class of 2010 down to 5.3 percent for the class of 2020.[24] The doctoral program of my department has become sufficiently competitive and complex that I was recently threatened with litigation after a candidate failed to gain admission. Not long ago, a postdoctoral competition in our field attracted two thousand applicants for a single position, and I've just submitted a letter of recommendation for a candidate pursuing another opportunity where last year's round saw one thousand competing for ten fellowships. I'm working on another letter for a competition that boasts of their "nationally recognized and extremely competitive program. We typically receive more than 750 applications per year and typically fund 4–5 new postdoctoral researchers per year." One of our current generation's leading spatial scientists pursued no fewer than forty-seven full-length academic job applications before landing a tenure-track position at a respectable but far-from-privileged institution. "In our plans for future professional action and in our advice to those in professional training," Ackerman warned, "we must think about these matters before it is much later."[25] Indeed. Then Ackerman warns what will happen if we, as human geographers, do not act quickly: "If we do not, others will cultivate our frontier, for that is the way of science. If we do, perhaps we may come closer to justifying Charles Darwin's words, '. . . that grand subject, that almost keystone of the Laws of creation, Geographical Distribution.'"[26] This is the last line of Ackerman's speech. It captures a crucial evolutionary theme in the geographical literature that connects nineteenth-century science and politics to a wide array of more recent pronouncements in our field—from Brian Berry's own 1980 AAG presidential address advocating "anticipatory evolution" through cybernetic "adaptive learning systems" to Harvey Miller's 2017 engagement with the evolutionary logics of the physicist/chemist Philip Ball to interpret recent GIScience advances and a "new social physics" in the "quest for geographic knowledge."[27] In the closing words of his speech in 1963, Ackerman is quoting a letter that Charles Darwin wrote to the botanist and explorer Joseph Dalton Hooker in 1845. The true significance of this citation becomes apparent only when we follow Ackerman's earlier injunction to "maintain a comprehensive view of the frontiers in the behavioral sciences," when we seek out "those investigators who pursue a systems approach" to diffusion, information, and the ways in which "social reality reflects the structure of the brain."[28] In Ackerman's day one of those frontiers—also funded by ONR—involved the search for the underlying systemic logics of the human capacity for language,[29] and that frontier still advances today thanks to the continued work of Noam Chomsky, who received the Atlas Award, the highest honor of the AAG, in Boston in April 2017. Hooker was one of Darwin's closest friends and confidants and had provided comments on advance proofs of the *Journal of Researches* that had been shared by the geologist Charles Lyell; Hooker and Lyell played the central roles in helping Darwin move quickly to avoid being scooped on the theory of natural selection by Alfred Russel Wallace, orchestrating the joint presentation of their research at the Linnaean Society in 1858. In turn, Darwin's *Origin*, the result of ideas he had begun formulating more than twenty years earlier, had been profoundly shaped by a blend of Thomas Malthus's influential *Population* essay and the most dynamic of the new disciplines that had emerged in the nineteenth century—geology, "freshly confected from mineral surveying, biblical chronology, and the study of fossils."[30] Just before the *Beagle* set sail in 1831, Captain Robert FitzRoy offered Darwin a welcome-aboard gift: the first volume of Lyell's *Principles of Geology*. Reading on the voyage, Darwin "had come to see the world with Lyellian eyes,"[31] governed by the principle of *incrementalism*, the cumulative effects of infinitesimally small changes over long periods of time. This is the part of Darwinian theory that is challenged by Chomsky's current work, coauthored with Robert C. Berwick, professor of computational linguistics in MIT's Laboratory for Information and Decision Systems and Institute for Data, Systems, and Society. Berwick and Chomsky write, "Darwin drank *Principles of Geology* neat. So do many origin of language theorists. Armed with Darwin and Lyell, they adopt a strong continuity assumption: like the eye and every other trait, language too must have evolved by 'numerous, successive, slight modifications.' . . . But is this strictly so?"[32]

Berwick and Chomsky's research provides evidence of rapid evolutionary transformation, a dynamic interaction effect between genetic inheritance and time-space sample selectivity in the variations within cohorts that exist prior to natural selection—eventually culminating in a remarkably rapid emergence of the "computational efficiency of language as a system of thought and understanding." Chomsky's challenge involves the speed and trajectory of evolution in the internal structure of the human brain and the capacity to translate the extremely limited empirical data provided by experience and sense perception into the nearly infinite diversity of "highly systematic," organized knowledge in the "mental creation of possible worlds."[33] Even now, nearly half a century since he articulated it so clearly in the famous 1971 Chomsky-Foucault debates (memorably described as an encounter of "two brains thinking simultaneously"), it remains an insurgent, minority view in linguistics, where an enduring, hegemonic dogma holds that language evolved slowly, incrementally, through social interaction as humans communicated with one another about externally observable phenomena.[34] Today's persistent dogma, of course, is a technologically enhanced perpetuation of that centuries-long Kantian cognoscitive powers referentialism doctrine, and we can see the inherently nonlinear dynamics of Ackerman's research frontiers when today's biolinguistics breakthroughs are "trace[d] back to the cognitive revolution of the seventeenth century, which in many ways foreshadows developments from the 1950s."[35]

Berwick and Chomsky help us to avoid looking just one way in a tunnel vision through the past and highlight a much more colorful, revealing parallax view in which internal cognoscitive powers are now coevolving with totalizing planetary God's-eye architectonic visions. Look back from Chomsky's current work through Ackerman's Darwinian geographical inspiration to Lyell's incrementalist geology, and then look forward from Lyell to William Morris Davis, who more than any one other figure built American geography in the late nineteenth century by finally secularizing geology—substituting evolutionary science and the idea of "inorganic natural selection" for the old teleological assumptions of German natural philosophy to create a "coherent geomorphological package" that he called "physiography."[36] Davis's evolutionary logic was in fact more neo-Lamarckian than Darwinian,[37] and hence it smuggled some dangerous assumptions about the intergenerational transmission of acquired characteristics—of landforms and of human cultures—into the theoretical core of the emerging discipline. This is the "deceptive simplicity" of environmental determinism, which Ackerman blames on the discipline's overreliance on connections with geology and history, but actually reflected a much wider panorama of political and intellectual struggles that can be traced all the way back to the birth of positivism itself. Victor Cousin, the conservative theological/political philosopher and nemesis of August Comte, espoused a "teleological metaphysics" suggesting that "national psyche could be read straight off topographic cartography: 'give me the map of a country . . . and I pledge myself to tell you, *a priori*, what the man of that country will be, and what part that country will play in history, not by accident, but of necessity.'"[38] It is not entirely coincidental that it was Cousin's doctrine of "interior observation," a fraudulent quack-psychology attempt to defend the crumbling castles of medieval theological political order against the onslaught of science discrediting Church doctrine in the 1830s and 1840s, that drove an enraged Comte to build such a strict scientism of *external* empiricist observation into his positivist philosophy.[39] Comte died the year before Darwinism began to fully undermine Lamarckian thought, and in the subsequent decades the speed differential between Lyell's geological evolution and the development of humanity became a central theme in American geographical thought: the neo-Lamarckian fusion of environmental and human transformation was "crucial to the project of carving out some cognitive space in academia for geography as a scholarly discipline."[40] This was achieved not only via Hartshorne-style bibliobiographies of specific theories,[41] but also through the Trevor Barnes geo-historiographies of Ackerman's "mental structuring" of important individuals. At Harvard, Davis's evolutionary geography was a powerful influence on Isaiah Bowman, who also absorbed enough pragmatism and New England neo-Kantian transcendentalism to develop what Neil Smith called an "evolutionary idealism"—"a belief in the hard-won, progressive development of ideas verified through a strict positivism."[42]

We cannot forget, however, that some of Bowman's early ideas never evolved. Despite the fact that the geologists had voted Ackerman summa cum laude and that the Davisian disciplinary paradigms of physiography and systematic geography had comprised two of his three doctoral examination subjects, Ackerman's post at Harvard was eliminated in no small part because of the enduring conservatism of the geologists' attitudes toward science. Abetted by Bowman's obsequious pandering to Conant the chemist, the geologists portrayed the idea of a human geography untethered from the physical sciences as illegitimate, "too easy." Hence it was the Washington think-tank branch of America's Cold War military-industrial complex that benefitted from Ackerman's 1945 vision of a planetary "totality of geography" understood perfectly by those in command of an army of technicians engaged in the great deal of mechanical work of the mind and eye. Look forward just two short decades to America's JFK New Frontier—"the frontier of the 1960s, the frontier of unknown opportunities and perils," beyond which "are uncharted problems of peace and war, unconquered problems of ignorance and prejudice"[43]—and Ackerman's militarized neo-Kantianism has become cybernation, research approaches toward understanding the process of human thought, and the study of the human brain in relation to a social reality governed by General Systems Theory.

Geographies of Individual and Collective Brains

Today, when Berwick and Chomsky remind us of the enduring Lyellian incrementalism that distorts evolutionary views of the past, the present provides abundant evidence of a sudden acceleration in evolutionary thought and practice that is shaping the newfound popularity of geography among billions of "citizen sensors" providing "volunteered geographic information."[44] That's because even as Berwick and Chomsky have assembled the best scientific evidence to narrow down the Hettner-Ackerman time-space coordinates at which "a slight rewiring of the brain" evolved to allow the human capacity for language,[45] the next generation of evolutionary innovators is competing to exploit and manipulate the frontiers of knowledge about that rewired brain and to connect these new versions of Victor Cousin's "interior observation" to the emerging global "collective brain" in some deeply troubling ways. When Chomsky spoke to analysts at Google in 2014, he was asked predictable Silicon Valley questions about the powers of natural language processing applied to the zettabytes of textual data from the globe-spanning cloud-based services that are now constantly interacting with billions of humans. Chomsky had to patiently remind the coders that "computational cognitive science" may yield a flood of patterns and correlations, but never any *understanding*. "That's not what understanding is," Chomsky emphasized; you get science not with an infinity of unguided observational quantifications, but with the rigor, discipline, mistakes, corrections, and creativity arising from the distinctively human ability to think, infer, and reason *beyond the limits* of empiricist observation. This is why Chomsky's work is so important for our understanding of the history of geography's quantitative revolution, and why Ackerman's concerns with Darwinian geography, cybernation, and the study of human thought help us understand today's proliferating paradoxes. For if language evolved in the spirit of the structuralist semiotician Ferdinand de Saussure, as "a social entity," "a storehouse of word images in the brains of a collectivity of individuals founded on a 'sort of contract,'"[46] then the algorithmic speed of big data observation and natural language processing would indeed yield genuine understanding—albeit with the consequence that, as Martin Rees points out, it will be the speedy machines and not us slow humans who will soon be doing most of that understanding.[47] But the evolutionary evidence is sharply at odds with this interpretation: rather than an externally focused communicative "referentialism" describing empirical observations, the capacity for language evolved as a byproduct of the ability to carry on long trains of thought, to construct "unbounded arrays of hierarchically structured expressions" from an ordered yet limited set of "mind-dependent, word-like atomic elements."[48]

In a world of networked social-media ecosystems now populated by billions, it literally makes all the difference in the world if language is not, in fact, a product of positivist external observation, but rather an "instrument of thought," a "system of meaning," an "inner mental tool" developed through the human capacity to make "infinite use of finite means."[49] The biolinguistics evidence produced and assembled by Berwick and Chomsky implies that there is a constant, inevitable, and inescapable contingency between (1) externally observable actions or expressions and (2) the brain's "conceptual-intentional interface" of human understanding, meaning, and thought.[50] Moreover, the systems of meaning that constitute *individual* human thought are not necessarily equivalent to the notion of an "inner mental tool": meaning is always simultaneously produced through the *interactions* between individual agency and collective, communicative processes. Such contingencies involve sweeping ontological implications that have shaped human social relations throughout the development of industrial capitalism,[51] ever since Marx, in the *Grundrisse*, satirized the way "Hegel fell into the illusion of conceiving the real as the product of thought concentrating itself, probing its own depths, and unfolding itself out of itself, by itself, whereas the method of rising from the abstract to the concrete is only the way in which thought appropriates the concrete, reproduces it as concrete in the mind."[52] But bad science and bad philosophy can be profitable and powerful politics. Computational cognitive sciences are all the rage, and as a rising share of human language is mediated through the "universe of self-replicating code" analyzed by the science historian George Dyson, we see a newer, faster, networked incarnation of the "cognitive Darwinism" that had developed at Chicago in the years before Derwent Whittlesey departed for Harvard.[53] The Kantian architectonic impulse is being rescaled in newly adaptive fusions of mind and world amid the "spatial swirl of capital and information"[54] that blends Norbert Wiener's cybernetic conception of humanity as "a communicative organism"[55] enmeshed with machines that have also become communicative organisms mediated by what Göran Therborn calls "computer-guided capital."[56] This contemporary "cognitive capitalism"[57] where "people and goods become complex entities that link and think"[58] is constituted through a dynamic reconfiguration of the relations between the *intrinsic* neurobiology of human brains and the *extrinsic* network characteristics connecting humans to one another (and to machines). At one scale, neuroscientists are using functional MRI as skull-scanning GPS techniques to map out the cognitive frontier, refining advertising and marketing by more precisely targeting the brain's neurological sites of Darwinian "meta-drives": survival, reproduction, kin selection, and reciprocal altruism.[59] There's a fast-growing new discipline of "evolutionary consumption" devoted to mapping the "evolved consuming instinct" of the "biological creature" now being called *homo consumericus.*

One of this new discipline's frontier pioneers, Gad Saad at the Molson School of Business at Concordia University, advises that fMRI and a Darwinized neuroimaging paradigm can be applied to the contents of cultural products—literary narratives, song lyrics, movie plotlines—to help us "identify universal themes that are indicative of a shared human nature that is invariant to time and place."[60] Neuroimaging shows that brains change, and all brains are in a continuous state of flux; this change must be understood dialectically, as intergenerational inherited precognitive wiring interacts with external stimuli. But the quantum shift in the scale and potential of cybernetic external stimuli is well illustrated by the millions enthralled by the stars of social media. Saad himself is a prime example, a cultural product he presents as the Gadfather, a scientist gleefully attacking leftist political correctness and building an audience of millions on his YouTube channel and talk show, *The Saad Truth.*[61] "I don't know if the Internet can handle such fabulousness!" Saad exclaimed when he introduced his guest in January 2016— Milo Yiannopoulos, the flamboyant provocateur at the center of the alt-right digital revanchism that was so instrumental in Donald Trump's campaign. Yiannopoulos, senior editor at *Breitbart,* had told Sky News that Trump would "make America fabulous again" and declared on Facebook, "I want to be Press Secretary in the Donald J. Trump White House. 'Today we bombed another third-world shithole. Also, how's my hair?'"[62] Yiannopoulos's conversation with Saad involved Milo's "Scientists Are Stupid" feature on *Breitbart*, which attacks high-profile science popularizers like Bill Nye the Science Guy and Neil deGrasse Tyson, whose media success, Yiannopoulos claims, depends entirely on compromising their science to conform with the "Ten Commandments of progressive liberalism." This is "not in keeping with the scientific method," Yiannopoulos complains, and Milo and Gad agree wholeheartedly that market-savvy liberal scientists suffer from a "lack of epistemological humility" on what we know, what we don't, and the importance of doubt in an era of rapid change. Yiannopoulos, wildly popular as a gay, conservative, and stylish Christian with a flair for the incendiary one-liner—he formulates his own environmental determinism concisely as "everywhere that doesn't have a strong Christian heritage is a bad place with fucked-up morals"[63]—satirizes Tyson as an "attention-seeking media troll" who ignores the fact that Isaac Newton "was a devout Christian who said the laws of nature were an expression of the divine will of God."[64] "Here we are on Earth, the perfect habitat for humanity with a million random variables somehow ending up in our favor," Milo thunders, and Tyson can't allow the benevolence of a higher power in the universe. But as the conversation with the Gadfather ranges from "leftist" scientists' arrogant attacks on social conservatives and Republicans, from Milo's critiques of the certainties of climate change and the politically correct strictures of the "social justice brigade" of "gender warriors" and "race hucksters" to the finer points of Milo's sexual preferences—Catholics have the best sex because it's forbidden!—the two dance coyly around their disagreements over the Darwinism of Saad's evolutionary psychology research. Saad has developed an entire paradigm of neuro-neoliberalism, and his evolutionary consumption interpretations of the U.S. presidential race— the heuristic power of visual cues calibrated for particular "evolutionary signatures" triggering voter-consumers[65]— point to the other, extrinsic scale of architectonic engineering, as cybernetically connected social lives mediate the quickened pace of mind-world manipulations. Cambridge Analytica, a spinoff of a British consulting firm and defense contractor involved in counterterrorism "psy ops" in Afghanistan as well as the United Kingdom's Brexit vote to leave the European Union, subsequently pivoted to help U.S. Republican candidates with micro-targeted "psychographic" political advertisements. Mining the results of hundreds of thousands of free "personality tests" on Facebook to calculate personalized OCEAN scores—ratings on the psychological traits of openness, conscientiousness, extraversion, agreeableness, and neuroticism—Cambridge Analytica claimed to have built a database of some three to five thousand data points on each of 230 million Americans. Nearly all are individually identifiable thanks to Facebook. Working for Trump in the 2016 general election campaign, one day in August the firm flooded the social network with a hundred thousand ad variations, what one analyst called "A/B testing on a biblical scale," refining its big data psychological profiling to precisely target dark posts for individual consumer-voters.[66]

New Cognitive Frontiers

All of these processes signify a newly accelerating dynamic in the development of geographic thought. It is not just that geographical theory incorporates evolutionary logics. Geographical ideas themselves—and indeed the entire discipline—are subject to evolutionary processes. This was Brian Berry's warning in his 1980 AAG presidential address: our field will survive only if we follow the economist Kenneth Boulding's conception of scientific theories as "mutations" in the "evolutionary ecosystem of mental species,"[67] and if we are able to sustain "[a] self-transforming profession adapted to the values of a self-transforming society—a creative geography actively involved in creating geography."[68] Ackerman's soaring address in 1963—with his retrospective on advances in geology and other physical sciences juxtaposed with a futuristic exhortation for geographers to deploy General Systems Theory to advance the frontiers of the science of human thought itself—was symptomatic of a broader transformation in scientific and political paradigms that would eventually remake America and much of the world. Enmeshed in a science think tank amid Washington's Cold War military-industrial complex, Ackerman was seeing the possibilities of a new kind of geography whose accelerated evolution was shifting conceptions and experiences of time, space, and locational variation. Ackerman was dazzled by the possibilities that such conceptions and experiences could be mediated, organized, and systematized through cybernetic reincarnations of the best of Kant, of Hettner, of Darwin. Yet this was a fragmentary, selective reincarnation, as Kant, Hettner, and Darwin were filtered through the influences of Hartshorne and the militarized knowledge hierarchies of the OSS. And even if Ackerman couldn't see it at the time, we now have the historical prerogative of contextualizing the wider politics giving rise to General Systems Theory.

This context involves the intertwined political-scientific

origins of cybernetics, systems theory, and neoliberalism. Von Bertalanffy's "Outline" begins with a section on "Parallel Evolution in Science" that proclaims the universal, "isomorphic" unity of physical and social systems that inspired Ackerman's cohort of scientific geographers. But right there on the second page, von Bertalanffy is citing Friedrich Hayek's *Road to Serfdom*: "In classical economic doctrine, society was considered as a sum of human individuals as social atoms. At present there is a tendency to consider a society, an economy, or a nation, as a whole which is super-ordinated to its parts. This conception is at the basis of all the various forms of collectivism, the consequences of which are often disastrous for the individual, and, in the history of our times, profoundly influence our lives. Civilizations appear, if not as superorganisms, as was maintained by Spengler, at least as superindividual units or systems, as expressed by Toynbee's conception of history."[69] Hayek is crucial here. We are living on the evolutionary frontiers of Ackerman's General Systems Theory and big data GIS implementations of Brian Berry's adaptive learning systems for anticipatory evolution. Frontier is now well beyond a simple metaphor: the same kind of Turner thesis conceptions of human nature, competition, and cultural evolution that rationalized imperial nation building in the twentieth century are now encoded in the algorithms of the free-market informational worlds envisioned by Hayek. And there's a horrific dark side to Berry and Wheeler's suggestion that communication and evolution will sort it all out, enabling the survival of the ideas best adapted to society's needs, to the aspirations of the next generation of scholars. This entire generational ontology of human scientific ambition—Ackerman, von Bertalanffy, Warren Weaver, Norbert Wiener, Brian Berry, and especially Friedrich von

Hayek—was conditioned by the cybernetic acceleration of machinic evolution in the post-1945 age of computers that mutated Marshall McLuhan's print-based *Gutenberg Galaxy* into the more dynamic, nonlinear digitalities of the *Medium Is the Message.*[70] The central illusion of this emergent informational society is that cognitive Darwinism ensures the survival of human intelligence—at least as we have typically defined that phrase, as we humans have come to understand and reflect on our abilities, our limitations, and our selves. Intelligence is indeed evolving faster in the network age, but there's absolutely no guarantee that it needs to be *human* intelligence. This, of course, is the vision of deep cogitations of a "machine-dominated culture" of Martin Rees, and of the Silicon Valley pioneers of the "singularity" of human consciousness uploaded to the networked global brain of a planetary internet.[71] And yet it's also the eureka dream that Hayek had almost a century ago, when he looked back across the human generations to the cognitive revolution of the Enlightenment. That sudden insight, in fact, was decisively *anti*-Enlightenment, a direct assault on the arrogance of rationality and human reason—the danger that Hayek called the "fatal conceit."

Hayek is today remembered as an economist, as the hero of all right-wing free marketeers from Barry Goldwater to Margaret Thatcher and Ronald Reagan to Ted Cruz and Paul Ryan. This part of his biography is true, but it's also incomplete and misleading. The worldview that Hayek regarded as his life's true contribution came not from economics but from "a sudden flash of insight" after his military service. "For a young man returning from World War I to enter the University of Vienna," Hayek recalled, his interests had "been drawn by those events from the family background of biology to social and philosophical issues."[72] Hayek's paternal grandfather taught natural science at a secondary school and wrote monographs in biology. Hayek's father, who became a medical doctor working for the Vienna ministry of health, was also a part-time lecturer in botany at the university, and the vast botanical collections he kept at home became young Friedrich's most enjoyable hobby. "My determination to become a scholar was certainly affected by the unsatisfied ambition of my father to become a university professor," Hayek reflected,[73] but in the catastrophic aftermath of the Great War, he took an interest in the power and danger of political and economic ideologies. At the university he was torn between psychology and economics, eventually opting for the latter as the safest and most lucrative career path; but the "fascination of physiological psychology," shaped by his enduring passions for biology and natural philosophy, remained at the core of all of his intellectual pursuits.[74]

As early as 1920, Hayek had developed a theory of human knowledge that took the neo-Kantian axiom—"to understand Kant is to go beyond him"[75]—to a dialectical extreme, creating an anti-Kantian fusion of Hume's empiricism and Darwinian evolution. In this formulation, all attributes of sense perception—and indeed "*all* mental qualities"—can be explained in terms of their place in a system of connections, and "the world of our mental qualities provide[s] us with an imperfect generic map with its own units existing only in that mental universe, yet serving to guide us more or less successfully in our environment."[76] At the precise moment when academic geography was elaborating Darwinian evolution into an all-encompassing environmental determinism at the scale of a world of colonial modernity, Hayek was doing something similar at the scale of the individual human mind.[77] "Mental events are a particular order of physical events within a subsystem of the physical world," Hayek concluded early on, "that relates the larger subsystem of the world that we call an organism . . . with the whole system so as to enable that organism to survive."[78] Empirical sense perceptions that give rise to human thought are not "carried" by the sensory fibers of neurons, Hayek was convinced, but are corelationally constituted through the "place of the impulse in a system of relations between all the neurons through which impulses are passed." Hence the human central nervous system must be understood as "an apparatus of multiple classification," as "a process of continuous and simultaneous classification and constant reclassification on many levels," applied not just to raw sensory inputs, but also to *all* "mental entities," including "emotions, concepts, images, drives, etc., that we find to occur in the mental universe."[79] This classification system means that the human mind is "a continuous stream of impulses," of external and internal stimuli and response, an input-output matrix "in which the state of the organism constantly changes from one set of dispositions to interpret and respond to what is acting upon it and in it," creating endless dynamic feedback loops between neurological processes and the organism's environment.[80]

None of this might sound like the *Road to Serfdom* manifesto that inspired Barry Goldwater and an entire generation of American conservatives (including the historian Harry Jaffa, who wrote Goldwater's famous "extremism in defense of liberty is no vice" proclamation to the 1964 Republican National Convention). But Hayek's psychology was and remains inseparable from his economics and politics. His philosophy joined the "mental universe" of neurological stimuli to the geopolitics of the world economy after Hayek, dissuaded by the poor financial prospects of biology or psychology, decided to pursue a career in economics. That journey—to the London School of Economics and epic battles with Keynes and the elite consensus on state economic planning—culminated in Hayek's architecture for a planetary environment conditioning the constant changes of an organism embedded in a continuous stream of impulses. Hayek's attack on socialism was driven by his conceptualization of the market as "a mechanism for communicating information." "In my own mind," Hayek explained to an audience of psychologists, there was absolutely no difference between the cognitive notes he eventually published as *The Sensory Order* in 1952 and the *Pure Theory of Capital* of 1941: mind and money evolved with one another. "I liked to compare" the endless stimulus-and-response circuitry and flow of "neural impulses, largely reflecting the structure of the world in which the central nervous system lives, to a stock of capital being nourished by inputs and giving a continuous stream of outputs—only fortunately, the stock of this capital cannot be used up."[81]

Such human cognitive capital attains infinity, however, only if we abandon any notion of human thought as embodied within individual human beings acting and learning in ways that we would describe in terms of consciousness, free will, or reason. Individual cognitive capacities are always and inescapably inferior to the collective consciousness created by the price signals of the market—the most powerful information processor ever created by the intergenerational selection dynamics of evolution. Markets gather dispersed pieces of partial, limited, and often contradictory knowledge, and operate through the price system to create a kind of mind: "How can the combination of fragments of knowledge existing in different minds bring about results which, if they were to be brought about deliberately, would require a knowledge on the part of the directing mind which no single person can possess?"[82] The price system *is* a mind, incorporating all of a society's changing needs and wants. Price signals drive the endless sifting and sorting of information about environment and resources, and of the ever-changing array of all possible configurations of production, distribution, consumption, and product innovation. The market develops its own "spontaneous order," and it always incorporates more and better information compared with the limited data available to any single individual or institution—even, and *especially*, the presumed experts of a socialist command economy or a managerial welfare state. *The market always knows best.* This is the "idea that swallowed the world,"[83] and it's also a full-fledged philosophical shift in the ontology of human evolution. It became the cognitive Darwinian source code for the exclusive club of "privatized, strategic, elite deliberation" in the "long-run *war of position* in the 'battle of ideas'" that Hayek launched at the Swiss resort of Mont Pèlerin sur Vevey in 1947.[84] Its corollary—that the only legitimate roles for the state involve the protection of property rights and the aggressive promotion of competition—was the central theme of Hayek's syllabus as he patiently tutored Thatcher as she rose to power: she once came to a seminar of the Centre for Policy Studies and pulled a ragged copy of Hayek's *Constitution of Liberty* from her handbag, slamming it down on the table: "*This* is what we believe!"[85] Thatcher and Reagan both credited Hayek as the inspiration for their free-market economic revolutions, and Lawrence Summers—he of the Clinton-era Committee to Save the World and later president of Harvard—has proclaimed that Hayek's theory of the price system as a mind is "the single most important thing to learn from an economics course today."[86]

There's a powerful American exceptionalism, however, in this conceptualization of the market as a mind. Hayek's American acolytes—from Goldwater to Reagan to Mike Pence—must always be careful to focus on the sound-bite free-market and antisocialist platitudes, because the "full Hayek" of human evolution is deeply offensive to the powerful Protestant Evangelical Christian constituents of the Republican coalition. Fewer than a fifth of U.S. adults concur that humans evolved with God playing no role in the process; the remainder are split evenly between the creationist "God created man in present form" versus the "intelligent design" version: "Human beings developed over millions of years, but God guided this process."[87] The theological intensification of American politics (almost exclusively among Republicans) is an important contextual shift, a literal devolution from the policy climate in which Ackerman could end a speech on scientific progress with a Darwinian hashtag. Hayek seems to have understood the need for trigger warnings for some of his right-wing snowflakes when he put the final touches on *The Fatal Conceit*, the first volume of his Collected Works, in the final year of the Reagan administration. Right on the first page, where he explains the power of evolutionary selection in the spontaneous emergence of traditions and moral practices that created differential advantages in survival and reproduction and yielded civilization, he cites scripture. "The unwitting, reluctant, even painful adoption of these practices kept these groups together, increased their access to valuable information of all sorts, and enabled them to be 'fruitful, and multiply, and replenish the earth, and subdue it' (Genesis 1:28)."[88] Three decades later, Trump Presidential Transition Team Executive Committee member Anthony Scaramucci appeared on CNN to defend a McCarthyite expedition for the names of disloyal Energy Department employees working on climate science: there's a lot we don't know, the Mooch explained to host Chris Cuomo, and "people have gotten things wrong throughout the 5,500 year history of our planet." People *always* get things wrong, Hayek emphasizes, because it is only "the extended order of human cooperation"—the market—that is capable of gathering and organizing information to yield solutions that enhance survival, prosperity, and reproduction. The fatal conceit is that everything about us—including our ability to communicate, our sense of right and wrong, our ability to learn, our very consciousness—is the product of a biological evolutionary path that can never be known, understood, or planned in advance. Whereas Hartshorne and Ackerman turned to Kant to stake a claim for geography alongside history, Hayek repeatedly lampoons Kantian idealism and reason, and indeed any kind of collective human institution or even monotheism itself: the spontaneous order arising from market relations is fully transcendent, and "*far surpasses the reach of our understanding, wishes and purposes*" as well as *any* single brain or will "(as, for example, that of an omniscient God)."[89] The fatal conceit is that we have evolved in ways that produce an activity we call "thinking," and this leads us to think that we can then start to play a role in that evolutionary process. We can't, Hayek insists. His injunction is absolute. The fatal conceit is "the idea that the ability to acquire skills stems from reason." We've turned the world upside down with the Kantian cognoscitive revolution, Hayek contends. "Mind is not a guide but a product of cultural evolution, and is based more on imitation than on insight or reason."[90] Hayek's philosophy is a doctrine of human ignorance and market intelligence, and in the United States it underwrites a bizarre survival-of-the-fittest economic policy ordained by God, where Jesus slashes welfare budgets and prefers low capital gains taxes. And there are other momentous implications. As Philip Mirowski has demonstrated, while neoliberal strategists did not develop the internet, they were the first to grasp the significance of its emergent combinatoric architecture for their political project. As Shannon-Weaver information theory, game theory, cybernetics, and operations research swept through the postwar sciences (especially economics), ideas of knowledge were replaced by quantified measures of volumes and probabilities in information, establishing the foundations for an "orthodox information economics" built on Hayek's conception of the market as the supreme evolutionary information processor.[91] Hence the axiomatic tautologies of Black Scholes options pricing and Alan Greenspan's insistence that financial bubbles can never be identified before the fact. The market, the powerful, transcendent processor that synthesizes all available information, always points the way to optimal allocation, innovation, and competition. Hence the stubborn "fail-and-flail forward" evolution of neoliberal policy: the only solution to catastrophic neoliberal marketization is the next neoliberal innovation.[92] Hence the informational semiotics of Republican media-market affirmation, from *Bedtime for Bonzo* Ronald Reagan to Jesse "The Body" Ventura to *Celebrity Apprentice* Donald Trump. "The Neoliberal Thought Collective is quite happy to have the masses mired in artificial ignorance," Mirowski and Nik-Khah wryly observe, "since that merely greases the wheels of The Market, that for which there is no greater intelligence."[93] Even, and *especially*, in the information society, *the market is always right*. Lies, rumors, interpretations, reputations, perceptions, facts—it's all sorted out in the market. "Alternative facts" become real when they find an audience in a competitive marketplace of ideas. At the moment, Donald Trump has 51.9 million Twitter followers. The conspiracy theorist Alex Jones, whom Trump praised for "an amazing reputation" in a full thirty-minute segment on Jones's *InfoWars* show only weeks before the 2016 Republican primaries began, racked up more than two billion views on his YouTube channel.

———

This is the new evolution of geographic thought. Charting a new frontier of cybernation in 1963, Ackerman described a General Systems Theory that emerged from the same kind of biological epistemology that inspired Hayek. And while Hartshorne and Ackerman are now gone and forgotten, Hayek built a doctrine of ignorance that is essentially the same authoritarian monster as Ackerman's 1945 militant neo-Kantianism. Hayek's rule of infinite marketization and monetization of information is now planetary—albeit variegated, incomplete, and sometimes contested. We must never forget that the production of variation, too, is central to the evolutionary process. Our networked world is defined by cosmopolitan diversity and difference. And yet it's also a world of ever-intensifying competition. How we think about that world, and the (in)justices of all those evolving differences, is one of the most urgent challenges in our present moment of geographic thought.

CHAPTER 6

Notes on Desk

Edward A. Ackerman, a brilliant small-town boy shaped by family tragedies that taught him what it meant to be truly all alone in the world, won a scholarship in 1930 that brought him to Harvard. That's where he fell in love with geography. Geography, at least in America, was also young and traumatized, struggling to gain recognition and legitimacy in the shifting hierarchies of academic knowledge and prestige. But in the 1940s, after all the academic accolades and all the years of dedicated service with America's most prominent geographers during the Second World War—"the best thing that [had] happened to geography since the birth of Strabo"[1]— Ackerman's promotion case triggered the death of geography at Harvard, which in turn pushed the dominoes everywhere else in the Ivy League. Ackerman's "extraordinary mental capacity" thus shaped geography from the outside, at the interface between academia and the applied imperatives of the Cold War military-industrial complex and its ancillary civilian components.

Ackerman's work at the Carnegie Institution of Washington helped shape the "newest 'new geography'"[2] of America's postwar worldview for a quarter century, in an era defined by interdisciplinary ambitions of modernist positivist certainty, scientific objectivity, and the flourishing computational advances of the digital age. Ackerman's journeys, always tracing paths between the centers and edges of America's role in the postwar global order, made his existence in Washington a busy, liminal affair in the space of flows of travel and communications. And so it is that Ackerman's Hägerstrandian life path leaves a thoroughly mundane imprint in the archives. One of the folders is hand labeled, "Notes on desk March 1973." Inside, among a page of scrawled notes and a typed manuscript on water development policies in countries of the United Nations Economic Commission for Europe, are a few telegrams on tractor-feed paper with cryptic header codes like `1-030728A060013 03/01/73 TLX UNATIONS NYK....` `MSC1355 REURLET 21 FEBRUARY EXTENSION DEADLINES ACCEPTED AS PROPOSED.`

Ackerman never met that extended deadline. He collapsed and died of a heart attack in Washington's National Airport. He was traveling, once again, on Carnegie business, in an informational world on the cusp of a geopolitically evolutionary mutation of General Systems Theory logics. Von Bertalanffy had died the year before, and the next year Hayek would be awarded the Prize in Economic Sciences for—as the Nobel Committee put it—his insight that "knowledge and information held by various actors can only be utilized in a decentralized market system with free competition and prices." By the time Ackerman died, the Pentagon Papers case over the classified knowledge and information disclosed by Daniel Ellsberg—and the Nixon administration's prior restraint of publication by the New York Times and the Washington Post—had already gone all the way to the Supreme Court, achieving a six-to-three victory. Yet by March 1973 Ellsberg himself was on trial in Los Angeles, facing six counts of espionage, six counts of theft, and one count of conspiracy. On the morning of Ackerman's death, the Times front page reported on the prosecution's attempt to defend the accuracy of politically manipulated Viet Cong troop strength statistics that the U.S. military had itself abandoned after the 1968 Tet Offensive. Elsewhere on the front page was a story noting that acting FBI director L. Patrick Gray had disclosed the Nixon administration's role in financing the undercover political espionage tactics of Donald Segretti. In a flurry of phone calls tracking down leads in the Watergate investigation, Carl Bernstein at the *Post* learned how Segretti's devious disruption of Democratic primary campaigns was known as "ratfucking." It reminded Bernstein of rumors he had heard about CIA practices known as "Mindfuck." Operatives were promised they'd be taken care of after Nixon's reelection. "How in hell are we going to be taken care of if no one knows what we are doing?" one of Segretti's dirty-trickster recruits asked. Segretti replied, "Nixon knows that something is being done. It's a typical deal. Don't-tell-me-anything-and-I-won't-know."[3] At almost exactly the same time that Bernstein was juggling phone calls in Washington, Vint Cerf at Stanford was working with Robert Kahn of the Information Processing Techniques office of the Pentagon's Defense Advanced Research Projects Agency, and sketched out the ideas of gateway architecture on the back of an envelope in a San Francisco hotel lobby. The system eventually came to be known as TCP/IP, the protocol of the contemporary internet. Later that year, University College London established the first international connection to the Pentagon's ARPAnet, via Norway's Royal Radar Establishment. But on the morning of Ackerman's passing, the microphones in the Oval Office were recording yet another conversation between Nixon and Henry Kissinger. Alexander Butterfield had not yet revealed the existence of Nixon's secret taping system—which several years earlier had captured Nixon's orders setting in motion covert operations to subvert Salvador Allende's democratically elected Socialist

government in Chile. Back in 1963, in the months when
Ackerman was working on his *Frontier* manifesto, he was also
busy commissioning a series of studies of the political stability
of Chile as a potential site for a major Carnegie-funded
Southern Observatory for astronomical research. Ackerman
had also contributed to a special section of the *Annals*,
"Critical Issues Concerning Geography in the Public Service,"
identifying the distinctive challenges of the Cold War—"a
peculiar sort of war, without any precedent in history."[4] "In
simple terms public policy is the way a controlling majority
or a controlling minority thinks before group action takes
place," Ackerman began, before considering the discipline's
historical and contemporary roles in military conflict,
domestic economic problems, and international development.
And then Ackerman reflected on the relations between real
and imagined geographies:

> We have used the phrase "cold war" not infrequently since
> its introduction several years ago. But the impression has
> settled over the general public that this phase in reality is
> a term for a kind of peace. It is not. We are not fighting
> for the physical possession of square miles of territory, or
> of strategic communities, or the destruction or defense of
> strategic military objectives in the sense of past wars. We
> are fighting for the adherence of nations and social groups
> to a way of life on which we believe the future of mankind
> depends. It is a war in which words like "coexistence" and
> "colonialism" are highly effective weapons. Indeed, we have
> seen whole empires disintegrate since the end of the last
> war under the impact of that one word "colonialism." It is
> a war in which the return of a single living man from a
> visit to the exosphere is publicly recognized as a significant

victory. Yet both sides intuitively recognize that the real significance of outer space exploration lies in the effect it will have on the minds of men in nations incapable of doing the same thing at this time. The real gains and losses can be geographically interpreted. It is in geographical space that the war eventually will be fought to a victory, loss, or stalemate. I believe that our public grasps this, and has illustrated it when a geographical loss has been suffered, like that of North Vietnam or Cuba.[5]

Later, the matter of words as effective weapons that could disintegrate empires came to define all the events unfolding in an unstable imperial cybernetic world in the last years of Ackerman's life. In Washington, the tapes rolled in the Oval Office—capturing, for example, Nixon chiding Kissinger for a lukewarm reaction to the idea of a nuclear response to a North Vietnam offensive: "The nuclear bomb, does that bother you? I just want you to think big, Henry, for Christsakes."[6] Meanwhile, as covert operations destroyed the "political stability" of the Chilean setting that Ackerman had examined for Southern Hemisphere views of the universe, another acolyte of Norbert Wiener's cybernetics and von Bertalanffy's systems theory—Stafford Beer, a visionary British operations research consultant—was sending frantic telexes from Manchester to Santiago in a desperate attempt to stem an unfolding public relations disaster. Right-wing newspapers in Chile were attacking Beer's Project Cybersyn, a cybernetic initiative to manage the Allende government's economic policies, as a sinister scheme for political manipulation.[7] Cybersyn, begun in 1971 at the invitation of Allende, was a new cybernetic approach to the organization of the social economy and used microwave links to connect mainframe computers linking most of Chilean industry together in a real-time system to achieve what Beer called "automatic statistical filtration of information." The system featured a Star Trek–style control room Beer described as an "ergonomically designed environment for decision."

There is a bittersweet significance to all of these connections and contingencies. By the time Ackerman died, the cybernetic hegemony envisioned in his 1963 *Frontier* speech had largely come to pass. If it failed to achieve its ambitions of universal acceptance across the physical and social sciences, at least it had become an aspirational guide for mainstream positivist scientific inquiry, a handy justification for investments in technological innovation, and a metaphorical guide for the everyday infrastructures of policy and the exercise of state power. From Ackerman's day to our current era of big data, Twitterbots, smart cities, and the Internet of Everything, from the transition from industrial modernity to postindustrial postmodernity, systems thinking of various kinds has endured as the universal popular unconscious. We continue to assume that more data equals more information, that more information yields more knowledge, and that advancing technological frontiers of information processing and distribution are inherently correlated with human capacities for coherent, rational thought, understanding, cooperation, and decision-making. And yet cyber-hegemony does not fulfill a systems ontology. A world of infinite data is entirely compatible with endless crisis, contradiction, uncertainty, and irrationality. Human information does not obey the laws of physics. In a cybernetic neoliberal world of Hayek's market-mind, the automated epistemologies of machine learning and artificial intelligence coevolve with reality television, weaponized social media, and alternative-facts post-truth politics. Segretti's Watergate-era "ratfucking" evolved into what Christopher Wylie, the Cambridge Analytica coder and whistleblower, calls "Steve Bannon's psychological warfare mindfuck tool."[8] Exponential growth in the production of data and information yields corresponding increases in what the mainstream economists describe as "asymmetric information"—theorized more critically and carefully by the science historian Robert Proctor as *agnotology*, the systematic production of ignorance.[9] Confusion and contradiction become profit opportunities and capital accumulation. Conspiracy becomes politics.

Systems ontology attains legitimacy only with a fully posthuman phase of evolution—in a closure of the epistemological space opened a century ago when human geography was created out of a discipline focused on the nonhuman phenomenon of geology (which itself had evolved from natural theology). Now we have a GPS smartphone world of popular, mobile geography in Hayek's natural theology of an algorithmically accelerated market as the most powerful information processor ever evolved in the universe: human geography without humans. My purpose in this book has been to connect some of the scattered fragments of our discipline's conceptual evolution, by exploring Ackerman's biography and written work—especially his eloquent 1963 *Frontier* presidential address. Part of the goal is to humanize Ackerman while raising deeply troubling questions about his work and his ideas, which were so deeply shaped by America's military-industrial complex and shaped an entire generation of quantitative revolutionaries. I want to talk with him in the same way Anne Buttimer spoke to Torsten Hägerstrand, and to show him how his frontier has evolved; of course we can't talk with Ackerman, so instead I'm pleading with *you* to love Ackerman, to hate some of his ideas, to read him—and most of all to know him as a *human geographer*. In the virtual and augmented realities of spaces and places in our present "Robotic Moment,"[10] the emphasis must be on the *human*.

"Bibliobiography," the word used by Richard Hartshorne to describe one of his many retrospectives on *The Nature of Geography*, was borrowed from John K. Wright. Wright had coined the term as he struggled to interpret the strange persistence of environmental determinism consolidated by a book published the year Ackerman was born: Ellen Churchill Semple's *Influences of Geographic Environment: On the Basis of Ratzel's System of Anthropogeography*.[11] Wright's concept isn't the same as a *biobibliography*; that's just a bibliography with a bit of biographical material mixed in. But a *bibliobiography* traces the birth, infancy, and maturity of a written work through its life and career. "A book responds to its environment by multiplying in number of copies more or less proportionately to its ability to make friends and interest people," Wright explained in a brilliant, sardonic, and sometimes vicious attack on deterministic thought. Judged by such criteria, Ackerman's written works were popular children who grew up to be forgotten loners. The clarion call of *Frontier*, General Systems Theory, was denounced the very same year by one of its cofounders: "Willingness to make a fool of oneself should be a requirement for admission to the Society of General Systems Research," declared Ken Boulding, who had been president of that society in 1957–1958. In geography, the framework became a pariah after Michael Chisholm reviewed Richard Chorley's attempts to reformulate the "inorganic natural selection" of William Morris Davis's "physiography" in GST terms and concluded that the extreme abstraction required rendered the approach "nothing but common sense. General Systems Theory seems to be an irrelevant distraction."[12] And yet this verdict is deceptively incomplete. Systems theory adapts and survives—the renamed International Society for the Systems Sciences met at Oregon State University in July 2018 under the theme "Innovation and Optimization in Nature and Design"—and it is now clear that systems theory emerged from a much wider coalescence of social physics monism, Shannon-Weaver information theory, Wiener-style cybernetics, and Hayekian evolutionary philosophies undergirding the development of a planetary Neoliberal Thought Collective.[13] It also reflected the sociopolitical environment of Cold War America.

Ackerman lived and worked amid all these influences, and so his ideas both reflected and shaped a crucial period in geography's quantitative revolution. Biobibliography and bibliobiography, therefore, are both essential in helping us come to terms with all the ambiguities, contradictions, and situated contingencies in the field we know as human geography. Ackerman was kind, thoughtful, and principled. And yet in his struggles to build his own career and to nurture an emergent geographical science, he was influenced by and contributed to a militarism premised on hierarchical control and the management of violence. His vision of geography on a behavioral science frontier that automates knowledge of human thought itself is now the essence of our everyday lives. Algorithmic aggregation through crowdsourcing, the adaptive auto-recommend interfaces augmented by artificial intelligence and a planetary noosphere of humans communicating through and with bots, the pattern-recognition techniques of state and corporate surveillance—all of these contemporary cybernetic advances represent the culmination of trends foreseen by Ackerman half a century ago. And all of them are quickly dehumanizing geographical thought and practice. With each passing generation, we venture further into the newest new frontiers of a Davisian and cognitive geomorphology, a militant neo-Kantianism where Hayek's environmental determinism of the mind erodes all nonmarket social relations in service to an ontology of ignorance in the face of a universal, omniscient, and unknowable spontaneous market order. We're living in the fatal paradox of Ackerman's frontier: Geography is more popular than ever before, as we networked billions carry our technologically advanced surveillance and navigation devices in our pockets everywhere we go. Silicon Valley is the next best thing to happen to geography since the birth of Strabo. But with all the automation, bots, and AI, it's no longer clear how many *human* geographers we'll need in this generation or the next. You and I—as human author and human reader—are able to meet only through the evolutionary human creations of language, the "instrument of thought" diagnosed by Noam Chomsky, and the mysterious time-spaces created in and through the geographies of the written word as explored by Michael Curry.[14] And yet the environment in which we encounter one another is now an unstable ecosystem of bots and algorithms, of search engine optimization (SEO) and automated application programming interfaces (APIs), of machine learning and natural language processing, of stylometry and adversarial stylometry. And so the "nexus of ideas-language-text-world" of the written work in Ackerman's era, the twilight of McLuhan's *Gutenberg Galaxy*, is now a new blend that combines text and multimedia with social structure in an evolving nexus that brings together the work as commodity, property, and meme.[15] The bibliobiography that Curry suggests for Henri Lefebvre's *Production of Space*— that Lefebvre's citations create places of dialogue and define the boundaries of a conceptual space, that "his own citations are entangled in a larger network of citations, compiled and disseminated by computers, and now over computer networks"[16]—now applies to us all. On the frontier encounters between savagery and civilization of Ackerman's frontier, as the continuous stream of impulses in our neurons interface with the continuous and simultaneous classification and reclassification algorithms of Hayek's planetary market-mind, we must work ever harder to find and defend a place to talk together, to learn from one another, as geographers, and as humans. These are the conversations that will be required for us to decide what kinds of humans we wish to become, and what kinds of human and nonhuman geographies we want to create in future generations.

Notes

PREFACE

1. John Bellamy Foster and Robert W. McChesney, "Surveillance Capitalism: Monopoly-Finance Capital, the Military-Industrial Complex, and the Digital Age," *Monthly Review*, July/August 2014, 1–31.
2. Donna J. Haraway, *Modest_Witness@Second_Millennium. FemaleMan©_Meets_OncoMouse™* (New York: Routledge, 1997).
3. George Dyson, *Turing's Cathedral: The Origins of the Digital Universe* (New York: Vintage, 2012).

CHAPTER 1

1. Richard Shearmur, "Debating Urban Technology: Technophiles, Luddites, and Citizens," *Urban Geography* 37, no. 6 (2016): 807–809, 807.
2. Daniel Ellsberg to Mrs. G. Philip Bauer, April 19, 1963, box 54, folder "April 1, 1963–April 29, 1963," Edward A. Ackerman Papers 1930–1973, American Heritage Center, University of Wyoming.
3. Edward A. Ackerman, "Where Is a Research Frontier?," *Annals of the Association of American Geographers* 53, no. 4 (1963): 429–440, 436.
4. Shearmur, "Debating Urban Technology," 808.
5. Andy Merrifield, *The New Urban Question* (London: Pluto Press, 2014).
6. Geoffrey J. Martin and Preston James, *All Possible Worlds: A History of Geographical Ideas*, 3rd ed. (New York: John Wiley, 1993), 433.
7. We can be quite certain that Ackerman indeed meant to write "removing individuality" by drawing inferences from the context of meanings in this section of "Where Is a Research Frontier?," and there is additional, archival evidence. The phrase is "removing

individuality" in the "short version" that presumably was the text for oral presentation; in the typescript of the full paper that was eventually published in the *Annals*, the passage has become "moving." Edward A. Ackerman, "Where Is a Research Frontier?" (1963), box 72, folder "Where Is a Research Frontier?," Ackerman Papers.

8. Norbert Wiener, *The Human Use of Human Beings: Cybernetics and Society* (Boston: Houghton Mifflin, 1950).

9. Edward Ullman to Edward Ackerman, August 18, 1944, box 47, folder "January 10, 1944–December 29, 1944," Ackerman Papers.

10. Ackerman, "Where Is a Research Frontier?," 435.

11. Geoffrey J. Martin, "The Nature of Geography and the Schaefer-Hartshorne Debate," in *Reflections on Richard Hartshorne's The Nature of Geography*, ed. J. Nicholas Entrikin and Stanley D. Brunn (Washington, D.C.: Association of American Geographers, 1989), 69–90, 74.

12. Peter R. Gould, "On *Reflections on Richard Hartshorne's The Nature of Geography*: Reflections Require a Mirror," *Annals of the Association of American Geographers* 81, no. 2 (1991): 328–334, esp. 330–331.

13. Robin Elisabeth Datel, "Taking a Moment to Bask in Our Past: Six Decades of APCG and AAG Presidential Addresses," *Yearbook of the Association of Pacific Coast Geographers* 62 (2000): 9–52, Table 6, 23.

14. Martin and James, *All Possible Worlds*, 433.

15. Trevor J. Barnes and Matthew Farish, "Between Regions: Science, Militarization, and American Geography from World War to Cold War," *Annals of the Association of American Geographers* 96, no. 4 (2006): 807–826.

16. George Dyson, "A Universe of Self-Replicating Code," *Edge*, March 26, 2012; George Dyson, *Turing's Cathedral: The Origins of the Digital Universe* (New York: Vintage, 2012).

17. Trevor J. Barnes, "Lives Lived and Lives Told: Biographies of Geography's Quantitative Revolution," *Environment and Planning D: Society and Space* 19 (2001): 409–429, 416.

18. Ron Johnston, "On Disciplinary History and Textbooks: or, Where Has Spatial Analysis Gone?," *Australian Geographical Studies* 38, no. 2 (2000): 125–137.

19. Radicati Group, *Email Statistics Report, 2016–2020* (Palo Alto, CA: Radicati Group, 2016), 2–3.

20. Susan Krashinsky, "Facebook Ups Effort to Protect Ad Space," *Globe & Mail*, August 10, 2016, B3.
21. Robert Bond, Christopher J. Fariss, Jason J. Jones, Adam D. I. Kramer, Cameron Marlow, James E. Settle, and James H. Fowler, "A 61-Million-Person Experiment in Social Influence and Political Mobilization," *Nature* 489 (2012): 295–298.
22. Adam D. I. Kramer, Jamie E. Guillory, and Jeffrey T. Hancock, "Experimental Evidence of Massive-Scale Emotional Contagion through Social Networks," *Proceedings of the National Academy of Sciences* 111, no. 24 (2014): 8788–8790.
23. And the partiality of these estimates also reflects the mundane frustrations of website paywalls and other access restrictions that prevent us from locating the firm's statistics on global mobile mapping applications. Estimates are from Aleks Buczkowski, "The US Mobile App Report—Google Maps App 64.5M Users, Apple Maps 42M," *Geo Awesomeness*, October 17, 2014.
24. John M. Crowley, "Denver Meeting of the Association of American Geographers," *Cahiers de Géographie du Québec* 8, no. 15 (1963): 103–107.
25. Rob Kitchin, *The Data Revolution: Big Data, Open Data, Data Infrastructures and Their Consequences* (Thousand Oaks, CA: Sage, 2014).
26. Michel Foucault, "What Is an Author?" (1969), in *The Essential Foucault*, ed. Paul Rabinouw and Nikolas Rose (New York: New Press, 2003), 377–391; Trevor J. Barnes, "Geo-historiographies," in *The Sage Handbook of Human Geography*, ed. Roger Lee et al. (London: Sage, 2014), 202–228, 204; Richard Hartshorne, "Notes toward a Bibliobiography of the Nature of Geography," *Annals of the Association of American Geographers* 69, no. 1 (1979): 63–76. Hartshorne borrows the word and concept from John K. Wright, who devised the idea for a 1961 paper tracing Ellen Semple's ideas on environmental determinism.
27. Daniel Ellsberg, *The Doomsday Machine: Confessions of a Nuclear War Planner* (New York: Bloomsbury, 2017), 55.
28. Ellsberg, *Doomsday Machine*, 55–56.
29. Ellsberg, *Doomsday Machine*, 65.
30. Ellsberg, *Doomsday Machine*, 65.
31. Trevor J. Barnes, "Taking the Pulse of the Dead: History and Philosophy of Geography, 2008–2009," *Progress in Human Geography* 34, no. 5 (2010): 668–677, 669.

CHAPTER 2

1.   David Harvey, *Explanation in Geography* (London: Edward Arnold, 1969), 105.
2.   Trevor J. Barnes, "Lives Lived and Lives Told: Biographies of Geography's Quantitative Revolution," *Environment and Planning D: Society and Space* 19 (2001): 409–429.
3.   For a biographical explication of the "life-lines" through which Torsten Hägerstrand came to develop the time-space paths for which he is now remembered, see the May 1979 interview conducted by Anne Buttimer. Ann Buttimer and Tom Mels, *By Northern Lights: On the Making of Geography in Sweden* (Aldershot: Ashgate, 2006), 107–121.
4.   See Torsten Hägerstrand's foreword to Buttimer and Mels, *By Northern Lights*, xi–xiv, xii.
5.   Again, Peter Gould provides the most candid and harsh assessment: "Apart from a few repetitive papers on location and the iron and steel industry . . . and another reporting on developments in political geography, . . . an article drawing on Hettner drawing on studies of the turn of the century, there is nothing except a stream of commentary and counter-commentary milking *The Nature* for over half a century. There is no substantive body of geographic *research* to evaluate." Peter R. Gould, "On *Reflections on Richard Hartshorne's* The Nature of Geography: Reflections Require a Mirror," *Annals of the Association of American Geographers* 81, no. 2 (1991): 331. Gould's assessment holds so long as we confine our attention to formal publications; as will become clear, I am also concerned with how human decisions and human relationships constitute the body of geographical research. In this sense, I am captivated by Gould's (331n4) polling of graduate students on Hartshorne's influence. One student responded, "He seems to be a ghost in the closet of previously traumatized senior faculty." Gould interprets this as a "generation gap": of Gould's 143 grad student respondents at Berkeley, Colorado, Indiana, North Carolina, Penn State, and Washington, 22 percent had never heard of *The Nature of Geography* and 57 percent had never read any of it. Every generation confronts its own gap. Peter Gould, who died in 2000, is now a ghost in the closet too, with an audience for the dance macabre that is even smaller than what he sampled for Hartshorne. But because I took classes with Gould when I was an undergraduate, he is a ghost in *my* closet, and because I have come to believe that "commentaries and counter-commentaries"

are human conversations worth paying attention to, I have forced you to read through this tedious footnote. But Gould's vicious vivisection of Hartshorne is important for our understanding of Ackerman's life and bibliobiography. If Gould were still with us he'd doubtless enjoy the Hägerstrandian, anti-Hartshorne implications of today's apocalyptic warnings of the latest incarnations of General Systems Theory—especially works like Nick Bostrom's *Superintelligence*. "Many of the points made in this book are probably wrong," Bostrom admits right in the preface, and then adds an endnote; page to the endnote, where he avers, "I don't know which ones." Bostrom's explanation of how he tried to write an easy-to-understand book that the *Sunday Telegraph* summarized as "a damn hard read" is quintessentially Hägerstrandian. "When writing, I had in mind as the target audience an earlier time-slice of myself," Bostrom explains, "and I tried to produce a kind of book that I would have enjoyed reading." Bostrom's warnings of an artificial intelligence explosion reveal just how far Ackerman's "research approaches" to "understanding the process of human thought itself" have evolved. Bostrom estimates a computational requirement of between 10^{31} and 10^{44} floating-point operations per second as a prerequisite for real-time simulation of the evolutionary path of all neuronal operations that have taken place in the history of life on Earth. Bostrom extrapolates such calculations across every domain that Ackerman had described in 1963—from the historic concentration of the world's superior cognitive capacities in Los Alamos for the Manhattan Project to the possibility of deploying Dyson spheres to achieve "whole brain emulations that live rich and happy lives while interacting with one another in virtual environments" across the billion-plus galaxies of Ackerman's admiration for an astronomy that "may stretch our minds most of all." And yet Gould also warned us about all of this, because what most enraged him about Hartshorne's *Nature* was the narrow, defensive disciplinary boundary-demarcation obsessions of "partitional thinking"—a mechanistic mind-set from Galileo and Newton that is now encoded into the taxonomies of an automated age, where the problems begin "not with the computers, but with us." Gould's warnings of the dangers of the coevolutionary acceleration of "the mechanistic world of things and the world of human beings"—and, in particular, the *thought* of humans— are a direct challenge to our present AI worlds of touch-screen convenience. "Machines That Think" have become dehumanized human "Thinks That Machine." Nick Bostrom, *Superintelligence:*

Paths, Dangers, Strategies (Oxford: Oxford University Press, 2014), viii, 261, viii, 102; Peter Gould, "Thinks That Machine," *Integrative Psychiatry* 3 (1985): 229–232, reprinted in Peter Gould, *Becoming a Geographer* (Syracuse: Syracuse University Press, 1999), 300–306, 304, 303.

6. The quote is from a letter Ackerman wrote in 1932 to W. G. Wendell, who endowed the scholarship enabling Ackerman's attendance at Harvard. Cited in Gilbert H. White, "Edward A. Ackerman, 1911–1973," *Annals of the Association of American Geographers* 64, no. 2 (1974): 297–309.

7. Donald J. Patton, "Obituary: Edward A. Ackerman," *Geographical Review* 64, no. (1974): 150–153, 151.

8. Harold S. Kemp to Preston James, May 1, 1941, box 47, folder "January 6, 1941–December 8, 1942," Edward A. Ackerman Papers 1930–1973, American Heritage Center, University of Wyoming.

9. Edward A. Ackerman, "Derwent Stainthorpe Whittlesey," *Geographical Review* 47, no. 3 (1957): 443–445. Whittlesey had to accept a demotion to assistant professor and a salary reduction in order to come to Harvard, where his first years were spent in subservience; he held and graded examinations for an absentee professor who spent most of the year traveling in Europe. Geoffrey J. Martin, "On Whittesey, Bowman, and Harvard," *Annals of the Association of American Geographers* 78, no. 1 (1988): 152–158.

10. Donald McLaughlin to Edward A. Ackerman, November 1, 1937, box 47, folder "September 6, 1934–December 6, 1939," Ackerman Papers.

11. Derwent Whittlesey to Donald H. McLaughlin, May 3, 1939, box 47, folder "September 6, 1934–December 6, 1939," Ackerman Papers.

12. Fred Lukermann, "*The Nature of Geography*: Post Hoc, Ergo Propter Hoc?," in *Reflections on Richard Hartshorne's* The Nature of Geography, ed. J. Nicholas Entrikin and Stanley D. Brunn (Washington, D.C.: Association of American Geographers, 1989), 53–68, 53.

13. The passage is from a 1978 history by Preston James and Geoffrey Martin, cited in Lukermann, "*The Nature of Geography*: Post Hoc," 54. The article in question was a 1937 *Annals* essay by John Leighly, which Lukermann (63) describes as "largely personal musings on the aesthetic nature of geographical literature dealing with the region, the 'vain dreams of a science of regions,' and the sterile ground of regional description." Lukermann (65) also emphasizes, however, that Hartshorne had been deeply concerned with

philosophy and methodological questions ever since his fieldwork in Germany in 1931–1932.

14. Lawrence Martin to Derwent Whittlesey, April 20, 1946, box 47, folder "January 2, 1946–September 24, 1946," Ackerman Papers.

15. Here again close reading yields editorial mysteries. Whittlesey may have been thinking of the "native hearth" of geographic methodology, but instead readers see "heath"—for which the first dictionary definition is "a tract of wasteland." Derwent Whittlesey, "A Foreword by the Editor," *Annals of the Association of American Geographers* 29, no. 3 (1939): 171–172. See also Geoffrey J. Martin, "The Nature of Geography and the Schaefer-Hartshorne Debate," in Entrikin and Brunn, *Reflections on Richard Hartshorne's* The Nature of Geography, 69–90.

16. See Geoffrey J. Martin, "Joe Russell Whitaker (1900–2000)," *Annals of the Association of American Geographers* 93, no. 1 (2003): 223–228.

17. Richard Hartshorne, "The Nature of Geography: A Critical Survey of Current Thought in the Light of the Past," *Annals of the Association of American Geographers* 29, no. 3 (1939): 173–412, and 29, no. 4 (1939): 413–658. The entire work is paginated independently; in the journal these numbers appear along with the sequence within the journal's own volume pagination.

18. Trevor Barnes and Jeremy Crampton, "Mapping Intelligence, American Geographers, and the Office of Strategic Services and GHQ/SCAP (Tokyo)," in *Reconstructing Conflict*, ed. Scott Kirsch and Colin Flint (Surrey: Ashgate, 2011), 227–251, esp. 239–240.

19. Edward Ullman to Edward Ackerman, July 7, 1944, box 47, folder "January 10, 1944–December 29, 1944," Ackerman Papers.

20. Barnes and Crampton, "Mapping Intelligence," 242–243.

21. Edward Ackerman to Charles C. Colby, August 13, 1945, box 47, "January 3, 1945–December 31, 1945," Ackerman Papers.

22. Ackerman to Colby, August 13, 1945.

23. Edward Ackerman to Lt. Col. H. G. Schenck, October 3, 1945, box 47, Ackerman Papers.

24. Edward A. Ackerman to Paul H. Buck, January 27, 1947, box 47, folder "May 1, 1946–December 20, 1947," Ackerman Papers.

25. Ackerman to Buck, January 27, 1947.

26. Ackerman to Buck, January 27, 1947.

27. White, "Ackerman," 300.

28. Trevor J. Barnes and Matthew Farish, "Between Regions: Science, Militarization, and American Geography from World War to Cold War," *Annals of the Association of American Geographers* 96, no. 4

(2006): 807–826, 808. See also Karl Butzer's assessment: "Almost 250 geographers (including most of the best) were engaged in some form of wartime government service in the Washington, D.C. area during 1942–45. . . . They were primarily employed in writing regional position papers or in cartography, and it seems inevitable that the regionalist and economic proclivities of the Midwestern–East Coast tradition would have then crystallized into a coherent paradigm, reinforced by close physical proximity and strong personal ties, that endured for many years as an elite club." Karl W. Butzer, "Hartshorne, Hettner, and *The Nature of Geography*," in Entrikin and Brunn, *Reflections on Richard Hartshorne's* The Nature of Geography, 50.

29. Hubert Schenck to Paul. H. Buck, June 11, 1947, box 47, Ackerman Papers.

30. Ackerman to Buck, January 27, 1947.

31. From a personal interview cited in Neil Smith, "'Academic War over the Field of Geography': The Elimination of Geography at Harvard, 1947–1951," *Annals of the Association of American Geographers* 77, no. 2 (1987): 155–172.

32. Hubert H. Schenck to Edward A. Ackerman, January 6, 1949, box 47, folder "January 6, 1949–December 31, 1949," Ackerman Papers.

33. White, "Ackerman," 299.

34. White, "Ackerman," 304.

35. The "superbrain" label comes from a January 1943 *New York Times* profile. See Larry Owens, "The Counterproductive Management of Science in the Second World War: Vannevar Bush and the Office of Scientific Research and Development," *Business History Review* 68 (1994): 515–576.

36. See Owens, "Counterproductive Management."

37. Dwight D. Eisenhower, "Farewell Address, January 17" (1961), reading copy, from Dwight D. Eisenhower Papers as President, Dwight D. Eisenhower Presidential Library, Abilene, KS, www .eisenhower.archives.gov/research/online_documents/farewell _address.html. Note that the quoted text comes from the reading copy and includes two edits in pencil by Eisenhower; "free ideas" has been edited to "new ideas," and "huge costs" has become "great costs." But in his broadcast speech, Ike went with the original word choices.

38. Thomas S. Kuhn, *The Structure of Scientific Revolutions* (Chicago: University of Chicago Press, 1962); Richard Peet, *Modern Geographical Thought* (Oxford: Blackwell, 1998), 23–26; George

Steinmetz, ed., *The Politics of Method in the Human Sciences: Positivism and Its Epistemological Others* (Durham, NC: Duke University Press, 2005).

39. Brian J. L. Berry and Edward J. Taafe, "The Location Theory Cluster" (Working Paper, NAS Committee on Geography, 1963), box 7, folder "NAS/NRC Committee on Geography, 1963–1966," Ackerman Papers.

40. Richard Hartshorne, *Perspective on the Nature of Geography* (Chicago: Rand McNally, 1959), 21, cited in Edward A. Ackerman, "Where Is a Research Frontier?," *Annals of the Association of American Geographers* 53, no. 4 (1963): 429–440, 434.

41. "The sciences differ enormously in their rates of progress.... Not all divisions of the behavioral sciences or the earth sciences offer channels for productive communication.... Cooperation must be selectively chosen. A good rule of thumb would be: where systems analysis techniques are understood and incorporated at the working face of the discipline, a collaborative definition of problem may profitably be sought." Ackerman, "Where Is a Research Frontier?," 438.

42. Ackerman, "Where Is a Research Frontier?," 435.

43. Ackerman, "Where Is a Research Frontier?," 437.

44. Kenneth E. Boulding, "General Systems Theory—The Skeleton of Science," *Management Science* 2 (1956): 197–208.

45. Wiener's "mental capacity" was perhaps even more "extraordinary" than what Whit claimed for Ackerman. Wiener earned his Harvard PhD at the age of nineteen. His *Cybernetics*, developed as a technical and philosophical outgrowth of his work in artillery control systems in World War II, was published in 1948. *The Human Use of Human Beings: Cybernetics and Society* (Boston: Houghton Mifflin, 1950), written for a much wider, nonspecialized audience, was published in 1950 and revised and updated for a 1954 edition.

46. See Wiener, *Human Use of Human Beings*, 7–12.

47. At one point, Hartshorne became absolutely enraged by the way an article published in the *Annals* had misrepresented his views and his interpretations of Kant, Humboldt, and Hettner. He wrote to the editor, "In whatever sense it is possible for a learned journal to commit a crime ... *The Annals* has committed a crime unparalleled in its history." Cited in Geoffrey G. Martin, "Nature of Geography and the Schaefer-Hartshorne Debate," 76.

48. Ackerman, "Where Is a Research Frontier?," 435. This is an astonishing assertion given Ackerman's (434) praise for Hartshorne's

decades of work "skillfully defending" the centrality of areal differentiation; it had been Fred Schaefer's 1953 attack on Hartshorne's
areal differentiation that had prompted the accusation that the
Annals had committed a crime.

49. Edward A. Ackerman, "Geographic Training, Wartime Research,
     and Immediate Professional Objectives," *Annals of the Association
     of American Geographers* 35, no. 4 (1945): 121–143, 122, 129.
50. To be sure, there are exceptions—exceptions that prove the
     rule. In a colloquium presentation in late September 1985 at the
     University of Minnesota, Fred Lukermann recounted how he
     had systematically waded through the check-out cards in the
     back covers of books in the university library—those cards with
     the names and dates indicating who had previously signed out
     a book—to infer how much Hartshorne had read of the sources
     he had cited. Daniel J. Hammel, just beginning graduate studies
     in the department, was struck by the clever possibilities of the
     research technique and soon headed to Wilson Library to see what
     famous patron names might be glimpsed in the cards for more
     current books; but soon the records were computerized, altering
     the methods and possibilities of bibliobiographical sleuthing. Neil
     Smith eventually got Hartshorne to admit that, in writing *The
     Nature of Geography*, he had really read only secondary sources
     and Kant's direct comments dealing with geography. Hartshorne
     began serious engagements with Kant's philosophical work only
     in 1970. Daniel J. Hammel, personal communication by email,
     October 13, 2016; Neil Smith, "Geography as Museum: Private
     History and Conservative Idealism in *The Nature of Geography*,"
     in Entrikin and Brunn, *Reflections on Richard Hartshorne's* The
     Nature of Geography, 91–120; Hartshorne's letter to Smith is cited
     on 100.
51. Vincent Berdoulay, "The Vidal-Durkheim Debate," in *Humanistic
     Geography: Prospects and Problems*, ed. David Ley and Marwyn S.
     Samuels (Chicago: Maroufa Press, 1978), 77–90, 88; see also
     Smith, "Geography as Museum," 96–100.
52. Robert B. Louden, "The Last Frontier: The Importance of Kant's
     Geography," *Environment & Planning D* 32 (2014): 450–465, 463.
53. See Robert C. Berwick and Noam Chomsky, *Why Only Us:
     Language and Evolution* (Cambridge, MA: MIT Press, 2016),
     85–86.
54. Joseph Stromberg, "Kantians with Cruise Missiles: The Highest
     Stage of 'Liberal' Imperialism," *Antiwar.com*, December 23, 2003.

55. Robert Curry, "Trump the 'Anti-Enlightenment Man' Revisited," *American Greatness*, March 16, 2017.

56. Joel Wainwright, *Geopiracy: Oaxaca, Militant Empiricism, and Geographical Thought* (New York: Palgrave Macmillan, 2013); see esp. xii–xiii and 72–73; see also the Book Review Symposium in *Human Geography* 7, no. 3 (2014).

57. Thomas E. Willey, *Back to Kant: The Revival of Kantianism in German Social and Historical Thought, 1860–1914* (Detroit: Wayne State University Press, 1978).

Chapter 3

1. Mark Rowlands, *The Body in Mind: Understanding Cognitive Processes* (Cambridge: Cambridge University Press, 2004), ix.

2. Edward A. Ackerman, "Where Is a Research Frontier?," *Annals of the Association of American Geographers* 53, no. 4 (1963): 429–440, 430.

3. Warren Weaver, "Science, Learning, and the Whole of Life" (address at the 70th Anniversary Convocation, Drexel Institute of Technology, December 1961), cited in Ackerman, "Where Is a Research Frontier?," 433.

4. Ackerman, "Where Is a Research Frontier?," 433, emphasis added.

5. David N. Livingstone, *The Geographical Tradition* (Oxford: Basil Blackwell, 1992), 114–115.

6. Ackerman, "Where Is a Research Frontier?," 435.

7. Marston Bates, "Summary Remarks: Process," in *Man's Role in Changing the Face of the Earth*, ed. William L. Thomas (Chicago: University of Chicago Press / Wenner-Grenn Foundation, 1956), 1139, cited in Ackerman, "Frontier," 429.

8. Ackerman, "Where Is a Research Frontier?," 429.

9. Trevor Barnes and Jeremy Crampton, "Mapping Intelligence, American Geographers, and the Office of Strategic Services and GHQ/SCAP (Tokyo)," in *Reconstructing Conflict*, ed. Scott Kirsch and Colin Flint (Surrey: Ashgate, 2011), 227–251, 233.

10. Hartshorne's words are from an undated memo in the NARA archives cited in Barnes and Crampton, "Mapping Intelligence," 233n10.

11. Barnes and Crampton, "Mapping Intelligence," 241.

12. Mina Rees, *Warren Weaver, 1894–1978* (Washington, D.C.: National Academy of Sciences, 1987), 513.

13. Warren Weaver, *Scene of Change: A Lifetime in American Science* (New York: Charles Scribner, 1970), 623.

14. Michel Valstar, Tobaias Baur, Angelo Cafaro, Alexandru Ghitulesco, Blaise Potard, Johannes Wagner, Elisabeth André, Laurent Durieu, Matthew Aylett, Soumia Dermouche, Catherine Pelachaud, Eduardo Coutinho, Björn Schuller, Yue Zhang, Dirk Heylan, Mariet Theune, and Jelte von Waterschoot, "Ask Alice: An Artificial Retrieval of Information Agent," in *Proceedings of the 18th ACM International Conference on Multimodal Interaction* (New York: ACM, 2016), 419–420, 419.

15. Ackerman, "Where Is a Research Frontier?," 429, 430.

16. Even the historical record of literatures and archives by which we seek to understand trends in scientific knowledge is contingent, embodied, and "fraught with political, social, and cultural anxieties." See Trevor J. Barnes, "Geo-historiographies," in *The Sage Handbook of Human Geography*, ed. Roger Lee et al. (London: Sage, 2014), 202–228, 224.

17. Neil Smith, "Geography as Museum: Private History and Conservative Idealism in *The Nature of Geography*," in *Reflections on Richard Hartshorne's* The Nature of Geography, ed. J. Nicholas Entrikin and Stanley D. Brunn (Washington, D.C.: Association of American Geographers, 1989), 91–120, 98–99.

18. Neil Smith, interview with Preston James, San Antonio, TX, April 27, 1982, cited in Neil Smith, *American Empire: Roosevelt's Geographer and the Prelude to Globalization* (Berkeley: University of California Press, 2003), 441. Compare with Ackerman's account of students, Whittlesey, and the opera: "For 25 years he owned a pair of Boston Symphony season tickets, so that he might share each concert or a season of concerts with a Harvard student. His generosity added greatly to the 'Harvard education' of a number of students." Edward A. Ackerman, "Derwent Stainthorpe Whittlesey," *Geographical Review* 47, no. 3 (1957): 443–445, 445.

19. Smith, *American Empire*, 442.

20. Saul B. Cohen, "Reflections on the Elimination of Geography at Harvard, 1947–1951," *Annals of the Association of American Geographers* 78, no. 1 (1988): 148–151, 148. Cohen (151) further recalls that "compared to voluminous lecture or reading notes in freshman History, Economics, and English, my freshman Geography I notes [with Kemp] were sparse and reflected a simple set of man-land relationship observations." "White shoe" is a descriptor that originated with casual buckskin shoes with red soles worn by men at Ivy League colleges; the term subsequently came

to be applied to law firms and financial companies, eventually acquiring a blend of connotations—elite, cautious, and conservative, but also effeminate and immature. See William Safire, "On Language: Gimme the Ol' White Shoe," *New York Times*, November 9, 1997.

21. According to two contemporaries in the department, Kemp was repeatedly verbally abusive to students and others in the university. "His ongoing patent mental problems resulted in some of the most vitriolic outbursts directed personally at individuals that we have heard in our lives." John P. Augelli and Donald J. Patton, "On 'Academic War over the Field of Geography,'" *Annals of the Association of American Geographers* 78, no. 1 (1988): 145–147, 145.

22. Carl Sauer, "Foreword to Historical Geography," *Annals of the Association of American Geographers* 31, no. 1 (1941): 1–27, 2. In "Where Is a Research Frontier?" (430), Ackerman offers a different interpretation, suggesting that the two fields geographers chose to ally with most closely—geology and history—"did not correct the predisposition of our scholars of the 'teens and early twenties to the deceptive simplicity of geographic determinism."

23. Smith, *American Empire*, 440.

24. Neil Smith, "'Academic War over the Field of Geography': The Elimination of Geography at Harvard, 1947–1951," *Annals of the Association of American Geographers* 77, no. 2 (1987): 155–172, 160.

25. Jean Gottman, interviewed by Smith, recounting a conversation with Bowman, cited in Smith, "Academic War," 162.

26. Smith, "Academic War," quotes from 156, 155.

27. Edward Ackerman to Paul Buck, August 1, 1948, box 47, folder "January 2, 1948–December 24, 1948," Edward A. Ackerman Papers 1930–1973, American Heritage Center, University of Wyoming.

28. Smith, "Academic War," 156.

29. Ackerman to Buck, August 1, 1948. The archived letter is handwritten, and one passage is illegible. "It can hardly be considered . . ." is corrected, with the first word crossed out; the final intended text may have been, "Larger actions can hardly be considered. . . ."

30. Brian J. L. Berry, "Paradigm Lost," *Urban Geography* 23 (2002): 441–445. See also the introduction to Brian J. L. Berry and James O. Wheeler, *Urban Geography in America, 1950–2000: Paradigms and Personalities* (New York: Routledge, 2005), xi–xix.

31. Edward A. Ackerman, "Where Is a Research Frontier?," *Annals of*

the Association of American Geographers 53, no. 4 (1963): 429–440, 439.

32. Ackerman, "Where Is a Research Frontier?," 440.
33. Ackerman, "Where Is a Research Frontier?," 439.
34. Hubert G. Schenck, *Publicity*, Memorandum NR 214.4, October 10, 1946, Ackerman Papers.
35. Rutherford Poats, "Expert Rules Japan Needs Birth Control," *Tokyo Stars & Stripes*, December 30, 1949, box 3, folder "Douglas MacArthur Material," Ackerman Papers.
36. Edward A. Ackerman, *Japanese Natural Resources* (1949), 528, draft excerpt, box 3, folder "Douglas MacArthur Material," Ackerman Papers.
37. Tiana Norgren, *Abortion before Birth Control: The Politics of Reproduction in Postwar Japan* (Princeton: Princeton University Press, 2001), 87.
38. Norgren, *Abortion before Birth Control*, 87–88.
39. "Allied Catholic Women's Club, Tokyo," copy of letter/memorandum, undated but ca. June 1949, box 3, folder "Douglas MacArthur Material," Ackerman Papers.
40. Norgren, *Abortion before Birth Control*.
41. Associated Press, "Vatican Assails Japan's Use of Birth Control," *New York Herald Tribune*, August 4, 1949, box 3, folder "Douglas MacArthur Material," Ackerman Papers.
42. Patrick O'Connor, "Women Hit Book Advocating Birth Control in Japan," *Boston Pilot*, January 15, 1950, box 3, folder "Douglas MacArthur Material," Ackerman Papers.
43. United Press, "Kaschmitter Attacks Birth Control Scheme," *Nippon Times*, January 3, 1950, box 3, folder "Douglas MacArthur Material," Ackerman Papers.
44. The book was eventually published by the University of Chicago Press.
45. Keyes Beech, "Wired to Chicago News, New York, Tokyo, 1200 hours, 8 Feb" (1950), box 3, folder "Douglas MacArthur Material," Ackerman Papers.
46. Beech, "Wired to Chicago."
47. Ackerman had accepted an appointment at the University of Chicago in 1948, but it's not exactly clear where he was in late 1949 and early 1950, given his multiple travels and contract assignments. He might have been in Washington serving on the President's Water Resources Policy Commission. See Gilbert H. White, "Edward A. Ackerman, 1911–1973," *Annals of the Association of American Geographers* 64, no. 2 (1974): 297–309, 299.

48. Edward Ackerman, *Statement of Edward Ackerman on Catholic Women's Club of Tokyo Objection to Report Japanese Natural Resources* (1950), box 3, folder "Douglas MacArthur Material," Ackerman Papers.

49. Edward Ackerman, Suggested Redraft of 528 (1949/1950), box 3, folder "Douglas MacArthur Material," Ackerman Papers.

50. United Press, "Mrs. Sanger Barred by MacArthur from Birth Control Talks in Japan," February 12, 1950, box 3, folder "Douglas MacArthur Material," Ackerman Papers.

51. Frank Luntz, *Words That Work: It's Not What You Say, It's What People Hear* (New York: Hachette, 2007).

52. Philip Mirowski and Edward Nik-Khah, *The Knowledge We Have Lost in Information: The History of Information in Modern Economics* (Oxford: Oxford University Press, 2017), 49.

CHAPTER 4

1. Edward A. Ackerman, "Where Is a Research Frontier?," *Annals of the Association of American Geographers* 53, no. 4 (1963): 429–440, 435, 436.

2. David Harvey, "The Urbanization of Consciousness" (1985), in *The Urban Experience* (Baltimore: Johns Hopkins University Press, 1989), 229–255.

3. Derek Gregory, "The Ideology of Control: Systems Theory and Geography," *Tijdschrift voor Economische en Sociale Geografie* 71, no. 6 (1980): 327–342, 332.

4. Gregory, "Ideology of Control," 341.

5. Allen J. Scott and Scott T. Roweis, "Urban Planning in Theory and Practice: A Reappraisal," *Environment and Planning A* 9 (1977): 1097–1119, 1114.

6. Michael R. Curry, "The Architectonic Impulse and the Reconceptualization of the Concrete in Contemporary Geography," in *Writing Worlds: Discourse, Text, and Metaphor in the Representation of Landscape*, ed. Trevor J. Barnes and James S. Duncan (New York: Routledge, 1992), 97–117, 97.

7. See also Kojin Karatani, *Architecture as Metaphor: Language, Number, Money*, trans. Sabu Kohsu (Cambridge, MA: MIT Press, 1995).

8. See, for example, Daniel Z. Sui, "Tobler's First Law of Geography: A Big Idea for a Small World?," *Annals of the Association of American Geographers* 94, no. 2 (2004): 269–277; Dawn J. Wright,

Michael F. Goodchild, and James D. Proctor, "Demystifying the Persistent Ambiguity of GIS as 'Tool' versus 'Science,'" *Annals of the Association of American Geographers* 87, no. 2 (1997): 346–362; Nadine Schuurman, "Trouble in the Heartland: GIS and Its Critics in the 1990s," *Progress in Human Geography* 24, no. 4 (2000): 569–590.

9. Richard Hartshorne, "The Nature of Geography: A Critical Survey of Current Thought in the Light of the Past," *Annals of the Association of American Geographers* 29, no. 3 (1939): 173–412, 310 [135].

10. Immanuel Kant, "Anthropology from a Pragmatic Point of View" (1800), in *The Cambridge Edition of the Works of Immanuel Kant in Translation: Anthropology, History, and Education*, ed. Gunter Zöller and Robert B. Louden (Cambridge: Cambridge University Press, 2007), 231–429, 233.

11. Robert B. Louden, "Translator's Introduction: Anthropology from a Pragmatic Point of View," in Zöller and Louden, *Cambridge Edition*, 227–230, 228.

12. The extensive note is Hartshorne, "Nature of Geography," 214–215 [38–39].

13. Quoted in Hartshorne, "Nature of Geography," 311 [135].

14. See Hartshorne, "Nature of Geography," 316 [140].

15. Edward A. Ackerman, "Geographic Training, Wartime Research, and Immediate Professional Objectives," *Annals of the Association of American Geographers* 35, no. 4 (1945): 121–143, 131.

16. Ackerman, "Geographic Training, Wartime Research," 132.

17. Lev Grossman, "Half the World Is Not Enough: Mark Zuckerberg's Plan to Get Every Human Online," *Time*, December 15, 2014, 31–41, 33.

18. Tim Craig, Missy Ryan, and Thomas Gibbons-Neff, "By Evening, a Hospital. By Morning, a War Zone," *Washington Post*, October 15, 2015.

19. Glenn Greenwald, *No Place to Hide: Edward Snowden, the NSA, and the U.S. Surveillance State* (New York: Metropolitan Books, 2014), 92–93.

20. Neil Smith, "Geography as Museum: Private History and Conservative Idealism in *The Nature of Geography*," in *Reflections on Richard Hartshorne's* The Nature of Geography, ed. J. Nicholas Entrikin and Stanley D. Brunn (Washington, D.C.: Association of American Geographers, 1989), 91–120, 98, 103; "only in our thoughts" is a direct quote from Hartshorne's "Nature of Geography."

21. Paul Richards, "Kant's Geography and Mental Maps," *Transactions of the Institute of British Geographers* 61 (1974): 1–16, 6.

22. David Harvey, "Cosmopolitanism in the *Anthropology* and *Geography*," in *Reading Kant's Geography*, ed. Stuart Elden and Eduardo Mendieta (Albany: State University of New York Press, 2011), 267–284.

23. Derek J. Gregory, "Killing over Kunduz," *Geographical Imaginations*, October 3, 2016, https://geographicalimaginations.com.

24. Cited in Jim Thatcher, "The Object of Mobile Spatial Data, the Subject in Mobile Spatial Research," *Big Data & Society*, July–December 2016, 1–7.

25. From the *Akademie Edition* of Kant's physical geography lectures, cited in Stuart Elden, "Reintroducing Kant's Geography," in Elden and Mendieta, *Reading Kant's Geography*, 1–15, 3. Kant saw both "inner" and "outer" sense domains of knowledge as inseparable dimensions for students' understanding; from 1772–1773 onward for several decades, Kant's teaching schedule was a two-course sequence—anthropology in the winter, physical geography in the summer.

26. Kant, "Preface to the First Edition" (1781), in *Kant's Critique of Pure Reason*, ed. J. M. D. Meiklejohn (London: G. Bell, 1910), xviii.

27. Gary Hatfield, "Introduction," in *Kant's Prologomena to Any Future Metaphysics, with Selections from the Critique of Pure Reason*, ed. and trans. Gary Hatfield (Cambridge: Cambridge University Press, 2004), ix–xxxiv, ix.

28. Hatfield, "Introduction," x.

29. Kant, *Critique of Pure Reason*, in Meiklejohn, *Kant's Critique of Pure Reason*, 1.

30. Smith, "Geography as Museum," 96.

31. Kant, *Critique of Pure Reason*, in Meiklejohn, *Kant's Critique of Pure Reason*, 8.

32. Smith, "Geography as Museum," 96.

33. *Kant's Inaugural Dissertation of 1770*, trans. William J. Eckoff (New York: Columbia College, 1894), 65.

34. From L. W. Beck's translation of the *Critique of Pure Reason*, cited in Richards, "Kant's Geography and Mental Maps," 4.

35. Richards, "Kant's Geography and Mental Maps," 7.

36. Collected Works, book IX, cited in Richards, "Kant's Geography and Mental Maps," 7.

37. Immanuel Kant, *Physical Geography* (1802), Rink edited

translation, in *The Cambridge Edition of the Works of Immanuel Kant in Translation: Natural Science*, ed. Eric Watkins (Cambridge: Cambridge University Press, 2012), 441–679, 446.

38. Kant, *Physical Geography*, 446.
39. Kant, *Physical Geography*, 446.
40. Houston Stewart Chamberlain, *Immanuel Kant: A Study and a Comparison with Goethe, Leonardo da Vinci, Bruno, Plato, and Descartes*, trans. Lord Redesdale (London: John Lane, 1914), 36.
41. Chamberlain, *Immanuel Kant*, 36.
42. Kant, *Physical Geography*, 447.
43. Peter R. Gould, *The Geographer at Work* (London: Routledge and Kegan Paul, 1985), 208.
44. H. Jesse Wheeler, "Evelyn Lord Pruitt, 1918–2000," *Annals of the Association of American Geographers* 96, no. 2 (2006): 432–439, esp. 436.
45. Ackerman, "Geographic Training, Wartime Research," 128.
46. Ackerman, "Geographic Training, Wartime Research," 139.
47. Ackerman, "Geographic Training, Wartime Research," 140.
48. Ackerman, "Geographic Training, Wartime Research," 137–138.
49. John Bohannon, "Mechanical Turk Upends Social Sciences," *Science* 352, no. 6291 (2016): 1263–1264, 1263.
50. Michel Foucault, "What Is an Author?" (1969), in *The Essential Foucault*, ed. Paul Rabinouw and Nikolas Rose (New York: New Press, 2003), 377–391; David N. Livingstone and D. T. Harrison, "Immanuel Kant, Subjectivism, and Human Geography: A Preliminary Investigation," *Transactions of the Institute of British Geographers* NS 6 (1981): 359–374.
51. Fred Lukermann, *Geography among the Sciences* (Minneapolis: Department of Geography, University of Minnesota, 1964), i.
52. Yi-Fu Tuan, *Who Am I? An Autobiography of Emotion, Mind, and Spirit* (Madison: University of Wisconsin Press, 1999), 96–97.
53. Trevor Barnes and Claudia Minca, "Nazi Spatial Theory: The Dark Geographies of Carl Schmitt and Walter Christaller," *Annals of the Association of American Geographers* 103 (2013): 669–683. See also Ken Olwig's historiography of the antecedents of Christaller's reactionary-modernist ideas of "isotropic *tabula rasa*" approaches to settlement geographies. Kenneth R. Olwig, "Transcendent Space, Reactionary-Modernism and the 'Diabolic' Sublime: Walter Christaller, Edgar Kant, and Geography's Origins as a Modern Spatial Science," *Geohumanities* 4, no. 1 (2018): 1–25, quote from 3.
54. Audrey Kobayashi, "Neoclassical Urban Theory and the Study

of Racism in Geography," *Urban Geography* 35, no. 5 (2014): 645–656, quote from 647.

55. Richard Shearmur, "Debating Urban Technology: Technophiles, Luddites, and Citizens," *Urban Geography* 37, no. 6 (2016): 807–809, 807.

56. Rob Kitchin, *The Data Revolution: Big Data, Open Data, Data Infrastructures and Their Consequences* (Thousand Oaks, CA: Sage, 2014), 21.

57. Don Mitchell, "Review of Joel Wainwright's *Geopiracy*," *Human Geography* 7, no. 3 (2014): 85–87.

58. Kant, *Physical Geography*, 446–447, emphasis original.

59. Kant, *Physical Geography*, 447.

60. Foucault, "What Is an Author?," 377.

61. See Paul C. Adams, *The Boundless Self: Communication in Physical and Virtual Spaces* (Syracuse, NY: Syracuse University Press, 2005).

62. Kant, *Physical Geography*, 447.

63. Michel Foucault, *The Order of Things* (1966; repr., London: Routledge, 2005), 347.

64. Foucault, *Order of Things*, 347.

65. Michael Curry, "The Idealist Dispute in Anglo-American Geography," *Canadian Geographer* 27, no. 1 (1982): 37–50.

66. Kant, "Anthropology from a Pragmatic Point of View," 232.

67. Kant, "Anthropology from a Pragmatic Point of View," 232.

68. Ian Hacking, *The Taming of Chance* (Cambridge: Cambridge University Press, 1990), 12.

69. Kant, "Anthropology from a Pragmatic Point of View," 231.

70. See Neil Smith, *American Empire: Roosevelt's Geographer and the Prelude to Globalization* (Berkeley: University of California Press, 2003), 44.

71. Joseph Urbas, "In Praise of Second-Rate French Philosophy: Reassessing Victor Cousin's Contribution to Transcendentalism," *Revue Française D'Études Américaines* 140, no. 3 (2014): 37–51, 38.

72. Robert C. Scharff, *Comte after Positivism* (Cambridge: Cambridge University Press, 1995).

73. Victor Cousin, *Lectures on the True, the Beautiful, and the Good* (1853), trans. O. W. Wight (New York: D. Appleton, 1890), 100, 101.

74. Bertrand Russell, *History of Western Philosophy, and Its Connection with Political and Social Circumstances from the Earliest Times to*

the Present Day (London: Allen & Unwin, 1946), 734, 739, emphasis added.

75. Ian Leslie, "The Scientists Who Make Apps Addictive," *1843: The Economist*, October/November 2016.

76. SFU Public Square, *Innovation: The Shock of the Possible, with Ray Kurzweil and Richard Florida* (Vancouver: Simon Fraser University, 2014). See also Kurzweil's "pattern recognition theory of mind": Ray Kurzweil, *How to Create a Mind: The Secret of Human Thought Revealed* (New York: Viking, 2012).

77. Rudolf A. Makkreel and Sebastian Luft, "Introduction," in *Neo-Kantianism in Contemporary Philosophy*, ed. Makkreel and Luft (Bloomington: Indiana University Press, 2010), 1–21; Deborah Withers, *Feminism, Digital Culture, and the Politics of Transmission: Theory, Practice, and Cultural Heritage* (London: Rowman & Littlefield, 2015).

78. Thatcher, "Object of Mobile Spatial Data," 2–3.

79. Thatcher, "Object of Mobile Spatial Data," 3.

80. Thatcher, "Object of Mobile Spatial Data," 3, 4.

81. Paul Feyerabend, *Against Method: Outline of an Anarchist Theory of Knowledge* (London: Verso, 1975), 70.

82. Kant, *Critique of Pure Reason*, in Meiklejohn, *Kant's Critique of Pure Reason*, 600.

83. Kant, *Critique of Pure Reason*, in Meiklejohn, *Kant's Critique of Pure Reason*, 599.

84. Slavoj Žižek, *Absolute Recoil: Towards a New Foundation of Dialectical Materialism* (London: Verso, 2014), 352.

85. Žižck, *Absolute Recoil*, 353.

86. Radicati Group, *Mobile Statistics Report, 2018–2022* (London: Radicati Group, 2018).

87. Stanley McChrystal, "The Illusion of Being Connected," *TEDx MidAtlantic*, April 20, 2014. See also Natasha O'Byrne, "The Quantified Self: Exploring the Ethics of e-Hancement Alongside the Rise of Planetary Urbanism," *Trail Six* 11 (2017): 74–85.

88. Jeremy Scahill, *The Assassination Complex: Inside the Government's Secret Drone Warfare Program* (New York: Simon & Schuster, 2016), 97.

89. Scahill, *Assassination Complex*, 97.

90. See R. G. Collingwood, *The Idea of History* (London: Oxford University Press, 1946), 96.

91. Collingwood, *Idea of History*, 131.

92. See Kate Connolly, "Protests over Terror Arrest of German Academic," *Guardian*, August 21, 2007; Anne Roth, "The

Interesting Case of Anne, the German Sociologist's Girlfriend,"
Wired, November 23, 2008. A petition calling for suspension of the
proceedings and restoration of freedom of research was signed by
thirty-two prominent scholars, including, inter alia, David Harvey,
Saskia Sassen, Neil Smith, Jennifer Wolch, Margit Mayer, Manuel
Aalbers, Mike Davis, Michael Dear, Roger Keil, Peter Marcuse,
Frances Fox Piven, and John Friedmann. See also Margit Mayer,
"Neil Smith: A Tribute from Berlin," *City* 16, no. 6 (2012): 689–691.

93. See, for example, Taylor Shelton, Matt Zook, and Mark Graham,
"The Technology of Religion: Mapping Religious Cyberscapes,"
Professional Geographer 64, no. 4 (2012): 602–617; Jeremy
Crampton, Mark Graham, and Ate Poorthuis, "Beyond the
Geotag: Situating 'Big Data' and Leveraging the Potential of the
Geoweb," *Cartography and Geographic Information Science* 40,
no. 2 (2013): 130–139; and Jim Thatcher, David O'Sullivan, and
Dhillon Mahmoudi, "Data Colonialism through Accumulation by
Dispossession: New Metaphors for Daily Data," *Environment &
Planning D* 34, no. 6 (2016): 990–1006.

94. Mark Graham, Ralph K. Straumann, and Bernie Hogan, "Digital
Divisions of Labor and Informational Magnetism: Mapping
Participation in Wikipedia," *Annals of the Association of American
Geographers* 105, no. 6 (2015): 1158–1178.

95. The case involved a University of Virginia–administered cruise
and two students: a twenty-year-old kinesiology major from
California Baptist University and a twenty-one-year-old Ohio
University student, taking a global studies class. See Susan Kinzie,
"An Education in the Dangers of Online Research," *Washington
Post*, August 10, 2008. News reports do not indicate the precise
technological means used to identify the alleged plagiarism, but
I strongly suspect the administrators relied on iParadigms, LLC's
Turnitin.com, the most widely used educational software application on the planet. Turnitin's core software was developed by
John Barrie, an undergraduate double major in neuroscience and
rhetoric at Berkeley who continued doctoral work in biophysics
and neurobiology to understand "how the brain encoded the sensory world into the neuro-world, how those patterns changed with
time, and ultimately how those patterns came together to form
our conscious representation of the world." Barrie's frustration
with rampant plagiarism when he served as a graduate teaching
assistant led him to adapt neurological software to analyze text.
See John Barrie, "Catching the Cheats: How Original," *Biochemist*
30, no. 6 (2008): 16–19.

96. Emma Teitel, "I Was a Plagiarist," *Macleans*, November 8, 2011.
97. Gilbert Ryle, *The Concept of Mind* (London: Hutchinson, 1949), 16.
98. Teitel, "I Was a Plagiarist."
99. See Thatcher, "Object of Mobile Spatial Data"; Crampton, Graham, and Poorthuis, "Beyond the Geotag"; Rob Kitchin and Martin Dodge, *Code/Space: Software and Everyday Life* (Cambridge, MA: MIT Press, 2011); Michael R. Curry, "The Digital Individual and the Private Realm," *Annals of the Association of American Geographers* 87 (1997): 681–699.
100. Ryle, *Concept of Mind*, 15–16, 13.
101. Sheldon J. Watts and Susan J. Watts, "On the Idealist Alternative in Geography and History," *Professional Geographer* 30, no. 2 (1978): 123–127, quote from 125.
102. Collingwood, *Idea of History*, 128.
103. Ludwig von Bertalanffy, "An Outline of General Systems Theory," *British Journal for the Philosophy of Science* 1, no. 2 (1950): 134–165, quote from 142.
104. Frederick Jackson Turner, "The Significance of the Frontier in American History," *Annual Report of the American Historical Association* (1893): 199–227, 200.

Chapter 5

1. Stephen D. N. Graham, "Software-Sorted Geographies," *Progress in Human Geography* 29, no. 5 (2005): 562–580; Jeremy Scahill, *The Assassination Complex: Inside the Government's Secret Drone Warfare Program* (New York: Simon & Schuster, 2016).
2. Zephoria, *The Top 20 Valuable Facebook Statistics—Updated May, 2018* (Sarasota, FL: Zephoria Digital Marketing, 2018).
3. Evgeny Morozov, *To Save Everything, Click Here* (New York: Public Affairs, 2013), 210.
4. Zephoria, *Top 20 Valuable Facebook Statistics*, citing an estimate published by CNN.
5. Imperva Encapsula, *2015 Bot Traffic Report* (Redwood Shores, CA: Imperva Encapsula, 2016).
6. Philip Howard, of the Oxford Internet Institute, cited in Adrienne Arsenault, "Partisan Twitter Bots Distorting U.S. Presidential Candidates' Popularity," *CBC News*, October 20, 2016.
7. Nicole Perlroth, "Internet Attack Spreads, Disrupting Major Websites," *New York Times*, October 21, 2016.

8. Political bot activity "reached an all-time high" in the 2016 cam-
    paign, and "not only did the pace of highly automated pro-Trump
    activity increase over time, but the gap between highly automated
    pro-Trump and pro-Clinton activity widened from 4:1 during
    the first debate to 5:1 by election day." Bence Kollanyi, Philip N.
    Howard, and Samuel C. Woolley, *Bots and Automation over
    Twitter during the U.S. Election* (COMPROP Data Memo, 2016.4;
    Oxford: Oxford Internet Institute, November 17, 2016).

9. Farhad Manjoo, "Social Media's Globe-Shaking Power," *New York
    Times*, November 16, 2016.

10. See Nadine Schuurman, "Tweet Me Your Talk: Geographical
    Learning and Knowledge Production 2.0," *Professional Geographer*
    65, no. 3 (2013): 369–377. Among the innovations of the schol-
    arship-industrial complex are the integration of Journal Citation
    Reports with Essential Science Indicators on the InCites platform;
    the integration promises an immersive "analytics-based discovery
    process" that includes dynamic, automated linkage to "third-party
    partner content for even deeper industry intelligence such as
    media monitoring and integration with ranking providers." Laura
    Gaze and Jen Breen, *Thomson Reuters Launches Game-Changing
    Enhancements to Its Flagship Research Analytics Platform Providing
    Unparalleled Scholarly Benchmarking & Analysis* (press release,
    Thomson Reuters, June 10, 2014).

11. See Jonah Peretti, "Capitalism and Schizophrenia: Contemporary
    Visual Culture and the Acceleration of Identity Formation/
    Dissolution," *Negations* 1 (Winter 1996), www.datawranglers.com
    /negations/issues/96w/96w_peretti.html.

12. Craig Silverman, "Viral Fake Election News Outperformed Real
    News on Facebook in Final Months of the U.S. Election," *Buzzfeed*,
    November 17, 2016.

13. See Gardiner Harris and Melissa Eddy, "Obama, with Angela
    Merkel in Berlin, Assails Spread of Fake News," *New York Times*,
    November 17, 2016.

14. Todd Witcher, "An Election Post-Mortem for 2016," *Huffington
    Post*, November 10, 2016.

15. Amy B. Wang, "Post-Truth Named 2016 Word of the Year by
    Oxford Dictionaries," *Washington Post*, November 16, 2016.

16. See Dana Priest, "NSA Growth Fueled by Need to Target
    Terrorists," *Washington Post*, July 21, 2013.

17. Ashley Parker and Maggie Haberman, "High in Tower, Trump
    Reads, Tweets, and Plans," *New York Times*, November 19, 2016.

18. Derek Hawkins, "Japanese American Internment Is 'Precedent'

for National Muslim Registry, Prominent Trump Backer Says," *Washington Post*, November 17, 2016.

19. This "mutated monist" geography now creates surreal, almost everyday Hägerstrandian intersections of the elusive "ground truth" component of the technological history of remote sensing and other elements of geographical information science. The *Washington Post* reporter Dana Priest describes meeting Lieutenant General Michael Flynn, a top intelligence chief who devised more effective tactical attack and information strategies in the Joint Special Operations Command (JSOC), at a 2008 cocktail party in Washington, D.C., hosted by the chairman of the Joint Chiefs of Staff. "Look at this!" Flynn exclaimed as he showed his cell phone to Priest. The screen showed a stream of updates from "tribal media outlets" in the restless frontiers of Pakistan's northwest provinces, chronicling the latest skirmishes and revenge killings. It was clear that while the rest of the U.S. military's bureaucracy had not yet understood how to make sense of the accelerated coevolution of social media, crowdsourcing, and "news," Flynn clearly understood how to read this "unfiltered information": "He was drawn to the little flecks of truth scattered on the ground." Later, as director of the Defense Intelligence Agency, Flynn became less interested in "flecks of truth" in favor of more extreme provocations—such as his false assertion that three-quarters of all new cell phones were bought by Africans—that led his staff to begin referring to "Flynn Facts." Removed from the directorship after eighteen months of disasters, Flynn jumped on the ex-military lecture circuit, and by the time he joined Trump's campaign as an advisor he was using Twitter to declare that "Fear of Muslims is RATIONAL," linking to fake stories about Islam seeking to enslave or exterminate 80 percent of humanity, and to spread false stories about NYPD discoveries of evidence of sex crimes in Hillary Clinton's emails. In hacked emails, Colin Powell described Flynn's trajectory as becoming "right-wing nutty," and as Priest summarized it, "The lifelong intelligence officer, who once valued tips gleaned from tribal reporters, has become a ready tweeter of hackneyed conspiracy theories." See John Pickles, ed., *Ground Truth: The Social Implications of Geographical Information Systems* (New York: Guilford, 1995); Dana Priest, "The Disruptive Career of Michael Flynn, Trump's National Security Advisor," *New Yorker*, November 23, 2016; Nicholas Kristof, "Trump Embarrasses Himself and Our Country," *New York Times*, November 19, 2016.

20. Glenn Greenwald, *No Place to Hide: Edward Snowden, the NSA,*

and the U.S. Surveillance State (New York: Metropolitan Books, 2014), 98–99.

21. See Michael Brennan, Sadia Afroz, and Rachel Greenstadt, "Adversarial Stylometry: Circumventing Authorship Recognition to Preserve Privacy and Anonymity," *ACM Transactions on Information and System Security* 15, no. 3 (2012): 12–22.

22. Edward A. Ackerman, "Where Is a Research Frontier?," *Annals of the Association of American Geographers* 53, no. 4 (1963): 429–440, 440.

23. H. Jesse Wheeler, "Evelyn Lord Pruitt, 1918–2000," *Annals of the Association of American Geographers* 96, no. 2 (2006): 432–439, 436.

24. Daphne C. Thomson, "Harvard Acceptance Rate Will Continue to Drop, Experts Say," *Harvard Crimson*, April 16, 2015. Part of the declining admissions rates for individual institutions can be attributed to the ease of applying to multiple institutions through the Common App, which was launched in 1998. Cybernetic convenience, however, has created a new social norm in which applicants are now routinely expected to apply to a dozen or more institutions.

25. Ackerman, "Where Is a Research Frontier?," 440.

26. Ackerman, "Where Is a Research Frontier?," 440

27. Brian J. L. Berry, "Creating Future Geographies," *Annals of the Association of American Geographers* 70, no. 4 (1980): 449–458, 456; Philip Ball, *Critical Mass: How One Thing Leads to Another* (New York: Macmillan, 2004); see also Philip Ball, "Gentrification Is a Natural Evolution," *Guardian*, November 19, 2014; Harvey J. Miller, "Geographic Information Science II: Mesogeography: Social Physics, GIScience, and the Quest for Geographic Knowledge," *Progress in Human Geography* 42, no. 4 (2018): 600–609.

28. Edward A. Ackerman, "Where Is a Research Frontier?," *Annals of the Association of American Geographers* 53, no. 4 (1963): 429–440, 438–439.

29. Noam Chomsky, "Three Models for the Description of Language," *IRE Transactions on Information Theory* 2, no. 3 (1956): 113–124; note, in particular, the funding acknowledgments for the research.

30. James A. Secord, "Introduction," in *Charles Darwin: Evolutionary Writings*, ed. Secord (Oxford: Oxford University Press, 2008), vii–xxxvii, quote from xii.

31. Secord, "Introduction," xiii.

32. Robert C. Berwick and Noam Chomsky, *Why Only Us: Language*

and Evolution (Cambridge, MA: MIT Press, 2016), 30, internal citation omitted (Darwin, 1859).

33. Berwick and Chomsky, *Why Only Us?*, 82, citing François Jacob.
34. Compare the 1971 Chomsky-Foucault debate, readily available on YouTube, with Chomsky's 2014 talk at Google; his portrayal of conventional linguistics as a "dogma" is from the 2014 talk.
35. Berwick and Chomsky, *Why Only Us?*, 82.
36. See David N. Livingstone, *The Geographical Tradition* (Oxford: Basil Blackwell, 1992), 202–204.
37. See Livingstone, *Geographical Tradition*, 205–212.
38. Quoted in David N. Livingstone, "Environmental Determinism," in *The Dictionary of Human Geography*, 4th ed., ed. R. J. Johnston, Derek Gregory, Geraldine Pratt, and Michael Watts (Malden, MA: Blackwell, 2000), 212–215, 213, internal citation omitted.
39. See Elvin Wyly, "Automated (Post)Positivism," *Urban Geography* 35, no. 5 (2014): 669–690, esp. 675.
40. Livingstone, *Geographical Tradition*, 189.
41. Environmental determinism had become deeply pervasive in Davis's generation. His own evolutionary thought is best expressed in a 1902 article on "systematic geography": "The relationship existing between the earth and its inhabitants must be explained under the broad principles of evolution. The earth with its lands and waters was not arranged for the convenience of its inhabitants; its inhabitants have had to learn, by more or less conscious experiment, to live upon the earth as they found it. . . . If the earth has not been expressly fitted to the convenience of its inhabitants, but if the inhabitants have had gradually to fit themselves to their slowly changing surroundings, how essential it is that we should study these surroundings minutely, with all the intelligence that has been awakened in the later days of man's history, in order to take the best advantage of them." William Morris Davis, "Systematic Geography," *Proceedings of the American Philosophical Society* 41 (1902): 235–259, 239.
42. Neil Smith, *American Empire: Roosevelt's Geographer and the Prelude to Globalization* (Berkeley: University of California Press, 2003), 43.
43. From Kennedy's nomination acceptance speech at the Democratic National Convention in Los Angeles, July 1960.
44. Michael F. Goodchild, "Citizens as Sensors: The World of Volunteered Geography," *GeoJournal* 69, no. 4 (2007): 211–221.
45. The change involves a set of fiber tracts in a connective circuit from Brodmann area 44, the dorsal-ventral pathway, and the

ventral pathway superior temporal cortex in the brains of our ancestors between eighty and two hundred thousand years ago in Eastern Africa. This corresponds to a range of five to six thousand human generations—not overnight, but also not "on the scale of geological eons." Berwick and Chomsky, *Why Only Us?*, 157.

46. Berwick and Chomsky, *Why Only Us?*, 95, citing Saussure.

47. Martin Rees, "Organic Intelligence Has No Long-Term Future," in *What to Think about Machines That Think*, ed. John Brockman (New York: Harper Perennial, 2015), 9–11, 11.

48. Berwick and Chomsky, *Why Only Us?*, 132.

49. Berwick and Chomsky, *Why Only Us?*, 102, 101, 164, 141.

50. Berwick and Chomsky, *Why Only Us?*, 164.

51. I am grateful to Audrey Kobayashi for valuable insights on these matters.

52. Karl Marx, *Grundrisse: Foundations of the Critique of Political Economy* (1857), trans. Martin Nicolaus (New York: Vintage, 1973), 101.

53. J. Nicholas Entrikin, "Robert Park's Human Ecology and Human Geography," *Annals of the Association of American Geographers* 70, no. 1 (1980): 43–58.

54. Neil Smith, "Spaces of Vulnerability; The Space of Flows and the Politics of Scale," *Critique of Anthropology* 16 (1996): 63–77, 71.

55. Norbert Wiener, *The Human Use of Human Beings: Cybernetics and Society* (Boston: Houghton Mifflin, 1950), 136.

56. Göran Therborn, "An Age of Progress?," *New Left Review* 99 (May/June 2016): 27–37, 36.

57. Yann Moulier Boutang, *Cognitive Capitalism*, trans. Ed Emery (Cambridge: Polity, 2011). See also Nigel Thrift, *Knowing Capitalism* (London: Sage, 2005).

58. Nigel Thrift, "Preface," in Boutang, *Cognitive Capitalism*, vi–x, viii.

59. Justin R. Garcia and Gad Saad, "Evolutionary Neuromarketing: Darwinizing the Neuroimaging Paradigm for Consumer Behavior," *Journal of Consumer Research* 7, nos. 4–5 (2008): 397–414.

60. Gad Saad, "Evolutionary Consumption," in *Wiley Encyclopedia of Management*, vol. 9, ed. Nick Lee and Andrew M. Farrell (Hoboken, NJ: Wiley Interscience, 2015), 1–5.

61. Saad's YouTube channel has more than 124,000 subscribers, and his videos have racked up more than 12.9 million views.

62. Posted on December 29, 2015.

63. On the *Joe Rogan Show*, October 2015.

64. Milo Yiannopoulos, "Scientists Who Are Actually Stupid: #1, Neil deGrasse Tyson," *Breitbart*, December 21, 2015.

65. See the interview of Saad in Kiki Sanford, "The Real Reason You're Voting for Clinton or Trump," *Nautilus*, October 21, 2016, http://nautil.us/blog/the-real-reason-youre-voting-for-clinton-or-trump. Among other insights, Saad explains that he knew the 1992 independent presidential candidate was unlikely to win because Ross Perot had "not won the genetic lottery" and failed to display the "signatures that matter from an evolutionary perspective."

66. McKenzie Funk, "The Secret Agenda of a Facebook Quiz," *New York Times*, November 19, 2016.

67. Berry, "Creating Future Geographies," citing Kenneth E. Boulding, "Science: Our Common Heritage," *Science* 297 (1980): 831–833.

68. Berry, "Creating Future Geographies," 455.

69. Ludwig von Bertalanffy, "An Outline of General Systems Theory," *British Journal for the Philosophy of Science* 1, no. 2 (1950): 134–165, 135.

70. Marshall McLuhan, *The Gutenberg Galaxy: The Making of Typographic Man* (Toronto: University of Toronto Press, 1962); Marshall McLuhan, *Understanding Media: The Extensions of Man* (New York: McGraw-Hill, 1964).

71. See Ray Kurzweil, *The Singularity Is Near: When Humans Transcend Biology* (New York: Viking Penguin, 2005); and Nick Bostrom, *Superintelligence: Paths, Dangers, Strategies* (Oxford: Oxford University Press, 2014).

72. F. A. Hayek, "The Sensory Order after 25 Years," in *Cognition and the Symbolic Processes*, vol. 2, ed. Walter B. Weimer and David S. Palermo (Hillsdale, NJ: Lawrence Erlbaum, 1982), 287–293, 287.

73. Quoted in Alan Ebenstein, *Friedrich Hayek: A Biography* (New York: St. Martin's, 2014), 8.

74. Hayek, "Sensory Order," 288.

75. Thomas E. Willey, *Back to Kant: The Revival of Kantianism in German Social and Historical Thought, 1860–1914* (Detroit: Wayne State University Press, 1978).

76. Hayek, "Sensory Order," 288.

77. Entrikin's history of the "cognitive Darwinism" and neo-Kantianism that shaped Robert Park's human ecology traces the influence of, among others, Heinrich Rickert; Entrikin cites a preface Hayek wrote for an English translation of Rickert's *Science and History*. The two dominant neo-Kantian influences on Robert Park's theoretical development (Rickert and Wilhelm Windelband) were

also fundamental in Hayek's epistemology of human knowledge. See Theo Papaioannou, *Reading Hayek in the 21st Century: A Critical Inquiry into His Political Thought* (Basingstoke: Palgrave Macmillan, 2012), 44–70, and Entrikin, "Robert Park's Human Ecology," 51–52.

78. Hayek, "Sensory Order," 288.
79. Hayek, "Sensory Order," 289.
80. Hayek, "Sensory Order," 291.
81. Hayek, "Sensory Order," 291.
82. Quoted in Stephen Metcalf, "Neoliberalism: The Idea That Swallowed the World," *Guardian*, August 18, 2017.
83. Metcalf, "Neoliberalism."
84. Jamie Peck, *Constructions of Neoliberal Reason* (Oxford: Oxford University Press, 2010), 49.
85. John Campbell, *The Iron Lady*, abridged by David Freeman (New York: Penguin, 2009), 93.
86. Quoted in Metcalf, "Neoliberalism."
87. Art Swift, "In U.S., Belief in Creationist View of Humans at New Low," *Gallup News*, May 22, 2017.
88. F. A. Hayek, *The Fatal Conceit: The Errors of Socialism*, ed. W. W. Bartley III (Chicago: University of Chicago Press, 1988), 6.
89. Hayek, *Fatal Conceit*, 72–73, emphasis original.
90. Hayek, *Fatal Conceit*, 21.
91. Philip Mirowski and Edward Nik-Khah, *The Knowledge We Have Lost in Information: The History of Information in Modern Economics* (Oxford: Oxford University Press, 2017), 46–50 and 66–72.
92. Peck, *Constructions of Neoliberal Reason*, 277.
93. Mirowski and Nik-Khah, *Knowledge We Have Lost*, 238.

Chapter 6

1. Kirk H. Stone, "Geography's Wartime Service," *Annals of the Association of American Geographers* 69, no. 1 (1979): 89–96, quote from 89.
2. Geoffrey J. Martin and Preston James, *All Possible Worlds: A History of Geographical Ideas*, 3rd ed. (New York: John Wiley, 1993), 433.
3. Carl Bernstein and Bob Woodward, *All the President's Men* (New York: Warner Books, 1974), 122.
4. Edward A. Ackerman, "Public Policy Issues for the Professional

Geographer," *Annals of the Association of American Geographers* 52, no. 3 (1962): 292–298, 296.

5. Ackerman, "Public Policy Issues," 296.

6. From a conversation on April 25, 1972, cited in Daniel Ellsberg, *The Doomsday Machine: Confessions of a Nuclear War Planner* (New York: Bloomsbury, 2017), 309.

7. See Eden Medina, *Cybernetic Revolutionaries: Technology and Politics in Allende's Chile* (Cambridge, MA: MIT Press, 2011), 185–187.

8. Carole Cadwalladr, "The Cambridge Analytica Files: 'I Made Steve Bannon's Psychological Warfare Tool': Meet the Data War Whistleblower," *Guardian*, March 18, 2018.

9. Robert N. Proctor and Londa Schiebinger, eds., *Agnotology: The Making and Unmaking of Ignorance* (Stanford, CA: Stanford University Press, 2008).

10. Sherry Turkle, *Alone Together: Why We Expect More from Technology and Less from Each Other* (New York: Basic Books, 2012).

11. John K. Wright, "Miss Semple's 'Influences of Geographic Environment': Notes toward a Bibliobiography," *Geographical Review* 42, no. 3 (1962): 346–361.

12. Michael Chisholm, "General Systems Theory and Geography," *Transactions of the Institute of British Geographers* 42 (December 1967): 45–52, 51.

13. See also Philip Mirowski, *Never Let a Serious Crisis Go to Waste: How Neoliberalism Survived the Financial Meltdown* (London: Verso, 2011).

14. Michael R. Curry, *The Work in the World: Geographical Practice and the Written Word* (Minneapolis: University of Minnesota Press, 1996).

15. See Curry, *Work in the World*, 185.

16. Curry, *Work in the World*, 201.

Index

CPSIA information can be obtained
at www.ICGtesting.com
Printed in the USA
FSHW021059120220
67034FS